John Henry Pratt

Scripture and Science not at Variance

With Remarks on the Historical Character, Plenary Inspiration... Sixth Edition

John Henry Pratt

Scripture and Science not at Variance

With Remarks on the Historical Character, Plenary Inspiration... Sixth Edition

ISBN/EAN: 9783337180331

Printed in Europe, USA, Canada, Australia, Japan

Cover: Foto ©ninafisch / pixelio.de

More available books at **www.hansebooks.com**

SCRIPTURE AND SCIENCE
NOT AT VARIANCE.

BY THE SAME AUTHOR.

ECLECTIC NOTES;

Or, Discussions on Religious Topics by the Members of the Eclectic Society of London, from A.D. 1798 to 1813.

Second Edition. Price 10s. 6d.

Some of the Members at this time were the Revs. Richard Cecil, John Venn, Thomas Scott, John Newton, Josiah Pratt.

PARAPHRASE OF THE REVELATION OF ST. JOHN,

According to the *Horæ Apocalypticæ* of the Rev. E. B. ELLIOTT.

Price 2s. 6d.

A TREATISE ON ATTRACTIONS, LA PLACE'S FUNCTIONS, AND THE FIGURE OF THE EARTH.

Fourth Edition. Price 6s. 6d.

The Second Edition of the author's work on 'Mechanical Philosophy' has been some years out of print, and has been succeeded by different treatises, by various authors, on its several subjects. The Treatise here advertised is an expansion of that part of the Mechanical Philosophy which treated of Attractions and the Figure of the Earth. The author has endeavoured to supply the want of a work on a subject of great importance and high interest—La Place's Co-efficients and Functions, and the calculation of the Figure of the Earth by means of his remarkable analysis. No student of the higher branches of Physical Astronomy should be ignorant of La Place's analysis and its results—'a calculus,' says Airy, 'the most singular in its nature, and the most powerful in its application, that has ever appeared.' He has also entered fully into the geodesic method of finding the Earth's figure, and introduced many propositions bearing upon that subject.

SCRIPTURE AND SCIENCE NOT AT VARIANCE;

WITH REMARKS ON

THE HISTORICAL CHARACTER, PLENARY INSPIRATION,
AND SURPASSING IMPORTANCE,
OF THE EARLIER CHAPTERS OF GENESIS.

BY

JOHN H. PRATT, M.A. F.R.S.
ARCHDEACON OF CALCUTTA,
AUTHOR OF 'THE MATHEMATICAL PRINCIPLES OF MECHANICAL PHILOSOPHY.'

SIXTH EDITION.

LONDON: STRANGEWAYS AND WALDEN, PRINTERS
28 Castle St. Leicester Sq.

PREFACE TO THE SIXTH EDITION.

It is now fifteen years since the first edition of this treatise was published. The last edition has been out of print more than a year; and I feel encouraged to send forth a sixth, especially as I have had various testimonies to the book having answered the end I had in view in writing it.

It was written in the first instance to meet the assertion made by the late Professor Baden Powell, that 'all geology is contrary to Scripture,' which I found was troubling many minds. I endeavoured to shape my argument in such a way as, not merely to be a reply to this mischief-working declaration, but to be a perpetual antidote to any other assertions of the kind which might emanate from the pen or lips of scientific or would-be-scientific men.

Since the first edition appeared, Professor Powell's *Order of Nature*, his 'Essay on Miracles' and Mr. Goodwin's on 'The Mosaic Cosmogony,' both in *Essays and Reviews*, Dr. Colenso's *Pentateuch*, Mr. Darwin's *Origin of Species*, and Sir Charles Lyell's *Antiquity of Man*, have been given to the world; and, with other minor productions, have been replied to in my successive editions, as far as they concern my argument.

And now, in the present edition, the following more recently published works are added: Professor Huxley's *Place of Man in Nature*, his *Lay Sermons, Lectures, and Addresses*, his and Professor Tyndall's Address and Lecture before the British Association at Liverpool, Sir John Lubbock's *Prehistoric Times* and his *Origin of Civilization and Primitive Condition of Man*, and Mr. Darwin's *Descent of Man*. The consequence is, that numerous additions have been inserted throughout, both in the text and in the notes. Several books not enumerated above are passed in review, or their suggestions made use of with acknowledgment. I regret that three excellent books—Dr. Beale's *Protoplasm*, his *Mystery of Life*, and Mr. St. George Mivart's *Genesis of Species*—came to my notice while these sheets were passing through the press, too late for me to make use of them. The part in which I treat on the Unity of the Human Race, or All Men of One Blood, has been much expanded. Three new illustrations of my leading principle have been added, and largely treated; viz. the Origin of Man, the Origin of Life, and Design as indicating creation. In short, no pains have been spared to bring the treatise down to the present state of Science and the controversies of the present time, as far as they appertain to the subject I have taken in hand.

The book has been thus increased by about half the amount of matter which the last edition contained; but the price remains as before.

JOHN H. PRATT.

Calcutta, 1871.

CONTENTS.

INTRODUCTION.

OBJECT AND PLAN OF THIS TREATISE 1

PART I.
THE HARMONY BETWEEN SCRIPTURE AND SCIENCE VINDICATED BY AN APPEAL TO THE HISTORY OF THE PAST.

CHAPTER
I.—EXAMPLES, FROM THE EARLIER HISTORY OF SCIENTIFIC DISCOVERY, IN WHICH SCRIPTURE HAS BEEN RELIEVED OF FALSE INTERPRETATIONS, AND THE HARMONY OF SCRIPTURE AND SCIENCE THEREBY RE-ESTABLISHED 14

    1. The Firmament. 2. Antipodes. 3. The Earth a Globe. 4. The Motion of the Earth.

II.—EXAMPLES, FROM THE LATER HISTORY OF SCIENCE, IN WHICH SCRIPTURE HAS NOT ONLY BEEN RELIEVED OF FALSE INTERPRETATIONS, BUT HAS HAD NEW LIGHT REFLECTED UPON IT BY THE DISCOVERIES OF SCIENCE 41

    1. The Antiquity of the Earth. 2. Creatures in Existence before the Six Days. 3. Existence of Light before the Six Days. 4. Death in the World before Adam's Fall. 5. Specific Centres of Creation. 6. No Known Traces of the Deluge. 7. The Deluge probably not over the whole Earth.

III.—EXAMPLES, IN WHICH SCIENCE HAS BEEN DELIVERED FROM THE FALSE CONCLUSIONS OF SOME OF ITS VOTARIES, AND THEREBY SHOWN TO BE IN ENTIRE AGREEMENT WITH SCRIPTURE . . 104

    1. All Men of one Blood. 2. Differences of Nations since the Flood. 3. Mankind originally of One Language. 4. Age of the Human Race according to Hindoo and Chinese Astronomy: 5. to Egyptian Antiquities: 6. to Nile Deposits: 7. to Flint Remains. 8. The Species of the Six Days' Creation distinct from pre-Adamite Species. 9. Origin of Species. 10. Origin of Man. 11. Origin of Life. 12. Uniformity of Nature. 13. Design. 14. Arithmetical Objections to the Pentateuch.

PART II.

THE HISTORICAL CHARACTER, PLENARY INSPIRATION, AND SURPASSING IMPORTANCE, OF THE FIRST ELEVEN CHAPTERS OF GENESIS.

CHAPTER PAGE

I.—THE HISTORICAL CHARACTER AND PLENARY INSPIRATION OF THE FIRST ELEVEN CHAPTERS OF GENESIS 335

    1. Our Lord and His Apostles regarded them as Historical documents. 2. This being the case, their Inspiration follows from the nature of their contents. 3. As they are original and not borrowed—Their freedom from Error

II.—THE SURPASSING IMPORTANCE OF THESE CHAPTERS 356

    1. They are of unrivalled Antiquity. 2. They tell us of the Origin of the World. 3. Of the Entrance of Evil into the World. 4. They explain the contradictions we see in Man. 5. They show the true basis of Physical Science, and the credibility of a Divine Incarnation. 6. They detect the essence of all successful Temptation. 7. They convey remarkable facts in History, the Institution of Marriage and of the Sabbath, the Deluge, the Confusion of Tongues, the apportioning of the Earth to the Nations, the Institution of Sacrifices. 8. They contain the germ of all Prophecy, the promise of the Seed of the Woman, and the prediction of the destinies of the descendants of Shem, Ham, and Japhet.

CONCLUSION.

NO NEW DISCOVERIES, HOWEVER STARTLING, NEED DISTURB OUR BELIEF IN THE PLENARY INSPIRATION OF SCRIPTURE, OR DAMP OUR ZEAL IN THE PURSUIT OF SCIENCE 369

SCRIPTURE AND SCIENCE NOT AT VARIANCE.

INTRODUCTION.

OBJECT AND PLAN OF THE PRESENT TREATISE.

THE assertion, not unfrequently made, that the discoveries of Science are opposed to the declarations of Holy Scripture, is as mischievous as it is false, because it tends both to call in question the Inspiration of the Sacred Volume and to throw discredit upon scientific pursuits.

Many who are predisposed to reject such a conclusion, from a general conviction that Scripture is the Word of God, are nevertheless at a loss for arguments to repel the charge. It is the object of the following pages to furnish such persons with a reply, in a concise and portable form. The treatise, therefore, is intentionally only a summary of arguments. To expand it, except by the addition of new illustrations, would defeat my design. A larger work would not find access where I hope this will. There are others also whose case it is here designed to meet—those who receive the Christian Revelation, but, under the influence of supposed difficulties brought to light by scientific discovery, are tempted to regard the Earlier Portion of the Sacred Volume as not inspired. It is possible that the unbeliever may find something in these pages to soften his prejudices: but his case is not here specially contemplated.

My treatise is, therefore, of the defensive kind. It is intended to show the inquiring how difficulties are to be met and objections removed. Some hesitate as to the expediency of putting such books into the hands of the young, thinking them calculated to engender doubts where they never existed, and to create the very scepticism which they were intended to rebut. There is some weight in this; and, no doubt, were the mind never likely in after-life to encounter the false views of sceptics, it might be far better to leave it untainted. If the young could always be fenced around by truth, till its principles became so thoroughly infused into their minds and hearts as to make error innocuous when they go out into the wide world, to leave them ignorant of the different forms of doubt and unbelief till circumstances force them upon their notice, might be the better course. But it is next to impossible to protect them, even when under the wisest guidance, from becoming acquainted with, if not imbibing, some of the mischief, which a refined scepticism—especially regarding the historical character and full inspiration of the Holy Scriptures—is spreading far and wide through the press and other channels. If the hesitation regarding the propriety of teaching these things to the young arise from a dislike to see old and *primâ facie* interpretations upset, such a course is most dangerous. By maintaining false and exploded interpretations as true, we are sowing in the minds of the young seeds of a future revulsion which is likely to injure them far more than the introduction of the new views at an earlier stage could possibly do. There can be no question that the safest course is conscientiously to teach the young the whole truth without reserve, not shrinking from stating in a plain and open manner the various objections and difficulties they will hear broached, explaining to them at the same time in what spirit and by what kind of argument they should be met.

The fact is, that sceptics and semi-sceptics are, unwittingly or not, undermining the faith of many in Scripture by subtle arguments drawn from the apparent contradictions between Scripture and Science. Against this it is necessary to provide an antidote: and the better fortified our youth are in their earlier days, the better prepared will they be to contend for the truth in after-life. It is not the Christian, but the worldly philosopher, who has raised these questions. But, having raised them, he forces the advocates of Scriptural truth to enter upon the contest, and to meet him on his own ground, that they may put a weapon of defence in the hands of those whose faith is in danger of being assailed.

I write for the protection and consolation of the faithful, under the attacks which Science, falsely-so-called, has brought against the Sacred Volume which is dear to them, and in which all their hopes of happiness are centred. There are excellent works which have the same end in view; but they take a different course. I will explain what I mean by an example.

INTRODUCTION.

Some advocates of Development, as the principle which has led to the present order of things, maintain, that creatures have their present habits, not because they were so created, and Divine Design is illustrated in the adaptation of their organs to those habits; but because no other habits could consist with such organs, that the organism has grown up in the natural course of things, in fact by natural law, and in cases where the habits were suitable to the organism, the organisms survived; in other cases they perished and disappeared. Now there are excellent treatises * which take up subjects of this kind on their own independent ground, not starting from the Scriptural side of the question; and the writers endeavour to show that an examination of the facts leads to an opposite conclusion; one which in the result coincides with Scripture statement. This requires a more lengthened treatment, and is very often successful. But there are instances in which this result cannot be thus absolutely attained; though nothing can be proved to the contrary. I take a different line in this treatise. I begin at the other end; and my aim is not to establish the truth of this or that theory which may be advanced, but to show that wherever any theory comes in conflict with Scripture rightly interpreted, it is the theory which is at fault, and not Scripture; if the theory does not touch upon Scripture, whatever it may be, I have in this treatise nothing to do with it. In adopting this line I feel it right to take the highest ground, and to

---

* I allude to such works as the Duke of Argyll's *Reign of Law* Dr. George Moore's (M.D.) *The First Man and his Place in Creation*, Mr. Gilbert Sutton's *Faith and Science*.

maintain it, till dislodged from it by argument and real facts. The Sacred Volume comes to us encompassed with evidence, external and internal, that it is the Written Word of God. This being the case, the most reasonable conclusion is, that it is free from error of every kind; for even where expressions are used which touch upon merely ordinary and natural things, it would be as easy for the inspiring Spirit to suggest to the minds of the writers words, not scientific words, but ordinary words, which would never be found at variance with fact, as words which, though they might at the time accord with current conceptions, would afterwards be found to be incorrect. Here, then, I take my stand: and I challenge Science—no, I will not so desecrate that honourable name by allowing even suspicion to attach to it, but I challenge Science falsely so called—to produce one instance in which the statements of Holy Scripture are proved to be wrong, except in as far as minor errors have crept in through the mistakes of the most careful copyists. I do not aim at reconciling Scripture and Science, though this is often the result of the investigation; but at demonstrating the fact which is involved in the title of my book, namely, that Scripture and Science are never at variance. This I do in the first part of my treatise, bringing together and examining all the examples I can think of, in which it has been alleged from time to time, that Scripture and Science are in irreconcilable conflict; and I show that further light or impartial examination has cleared up the difficulty. From this I argue, that it is in the highest degree *unphilosophical*, wherever new difficulties arise in these days of discovery, to doubt that these also will be cleared up as light and knowledge advance. The experience of the past should encourage us fearlessly to carry our investigations into the phenomena of nature, fully persuaded no real discrepancy can ever be in the end established. The above may be regarded as a negative argument.

In the Second Part I enter upon an examination of the character and contents of the earlier portion of the Book of Genesis; as it is in this portion of the Sacred Volume that the seeds of strife between Scripture and Science are supposed chiefly to lie. By what I cannot but regard as an unanswerable proof of the historical character and plenary inspiration of these Early Chapters, and by a reference to their important bearing in various eminent particulars, I establish a positive argument for their inspiration, and show that under these circumstances it is *impossible* that Scripture can, when rightly interpreted, be at variance with the Works of the Divine Hand; and that therefore, if difficulties remain at any time not cleared up, they must arise from our ignorance, or from hasty interpretation either of the phenomena before us or of the language of the Sacred Record.

The results of this investigation are then summed up, and the conclusion drawn,—that no new discoveries, however startling they may appear at first, need disturb our belief in the Plenary Inspiration of the Sacred Volume, or damp our ardour in the pursuit of Science.

It will be seen from the above sketch, that it is not necessary for the validity of my argument that every instance of apparent discrepancy between Scripture and Science shall have met with an explanation. It requires only, that so many instances of the successful removal of difficulties, which at one time appeared to be insurmountable, should be adduced, as to assure the mind under new perplexities, that there is every reason to believe that in time these also will vanish. The primary object of the treatise is, not to solve present difficulties, but to create confidence in the mind, while in perplexity regarding them, that all will in the end be right, and that the harmony of Scripture and Science *cannot* really be broken, though it may for a time seem to be disturbed. In point of fact, however, I know of no alleged or apparent discrepancy between Scripture and Science which cannot be met by a decisive or at least satisfactory answer. The chief examples I have brought together in the following pages, and have made them the groundwork of my argument. Had I known of any existing unanswered difficulty, I should now have brought it forward as an illustration of the use of my principle. If, for example, the sweeping announcement of M. Bunsen and Mr. Leonard Horner, that the age of the human race is many thousands of years older than the Scripture narrative makes it, or the same from the pen of Sir Charles Lyell, in his work *On the Antiquity of Man*, or the hypothesis that the descent of the human race is not to be reckoned from Adam, but, as Mr. Darwin and Professor Huxley conjecture, from some primitive monad, the progenitor of all plants and animals, could not yet be met, I should have produced it,—not, as in the present edition, doing homage to my argument, but as an example of the principle I have set forth, that we should wait, fortified by the experience of the past, and by an immovable belief in the inspiration of Holy Scripture, and feel assured that time would turn objections into proofs, and discrepancy into harmony.

The result, therefore, of my treatise, beyond its direct object to inspire confidence for the future, brings out this,—that, notwithstanding the assertions of certain writers, nothing has been produced and established which is really contradictory to the statements of Holy Scripture. Guesses and crude speculations have been substituted for facts, and what has been in these instances called Science is not worthy of the name. Deeply conscious of the goodness and truth of our cause, we can afford to smile at, and forgive, such rough and unpolished shafts as the following, aimed at us, who maintain and defend the integrity and inspiration of God's Holy Word:—'Extinguished theologians lie about the cradle of every science as the strangled snakes beside that of Hercules. . . . Orthodoxy is the Bourbon of the world of thought. It learns not, neither can it forget. . . . Philosophers may, now and then, be stirred to momentary wrath by the unnecessary obstacles with which the ignorant, or the malicious, encumber, if they cannot bar, the difficult path [of the progress of discovery].'

We are convinced indeed, that, in the annals of Science, there is no class which stands more prominently forward in supplying leaders of scientific thought and of scientific discovery, than the defenders of the Sacred Volume, against whom these bitter words are uttered. We cannot, however, with the same composure, overlook and forget the ignorance and irreverence which shock our ears by representing any part of the Holy Scriptures as merely 'the imaginations current among the rude inhabitants of Palestine.'

We pity from our hearts the men who regard what they call 'the cosmogony of the semi-barbarous Hebrew,' as 'the incubus of the philosopher and the opprobrium of the orthodox;'* for they deeply injure their own minds by holding such views, help to bewilder and mislead the young and the enquiring by throwing among them these sentiments broad-cast in their writings; and they cut themselves off from enjoyments, intellectual and spiritual, of which we would see them participate as well as ourselves.

* These quotations are from a book published in 1870.

PART I.

THE HARMONY BETWEEN SCRIPTURE AND SCIENCE VINDICATED BY AN APPEAL TO THE HISTORY OF THE PAST.

THE Book of Nature and the Word of God emanate from the same infallible Author, and therefore cannot be at variance. But man is a fallible interpreter; and by mistaking one or both of these Divine Records, he forces them too often into unnatural conflict.

Reason, when combined with a humble mind and a patient spirit, is man's highest endowment. By it he can scale the heavens, and unravel the mysterious ties which unite matter to matter in all its combinations; and can trace the secret and silent operation of its laws. Thus furnished, he can weigh and appreciate the claims of truth, as revealed from heaven or produced from the evolutions of the human mind; and can reject the evil and choose the good. But, deprived of these valuable accessories, this noble gift is converted into a snare, and too often hurries him to conclusions from which he is afterwards compelled to retrace his steps.

It is my intention to bring together in this First Part of my treatise a number of Examples, gathered from the history of Science, which show how needless are the fears entertained at the present day by many excellent persons in their holy jealousy for the Sacred Volume, in which their highest hopes are centred; as it has already, in so many instances, triumphantly emerged from conflicts, as severe as any in which it may now or hereafter be engaged.

In some instances, positive errors in the interpretation of the phenomena of nature, and in others ignorance of the facts of nature, have led to the imposing upon Scripture a meaning, which the correction of these errors on the one hand, and the discovery of new facts on the other, have proved to be false. As true Science has advanced, Scripture, so far as it touches upon natural phenomena, has received new illustrations. False interpretations have been detected and corrected. The language of Scripture has been found to be in no case opposed to truth. It in no case stoops to the errors and prejudices of men, even in things natural, although it adopts the language of men and its usages. It speaks on such matters as man would speak to man in every-day life, in the times of greatest scientific light. It selects no particular epoch of discovery for the choice of its phraseology; but it speaks, as the most scientific amongst us speak, in the ordinary intercourse of life, in conformity with the usages of language—namely, according to appearances.

The Examples, above referred to, I shall class under three heads. The first class arose from the progress of discovery sweeping away long-standing notions regarding the nature of the canopy above us, the existence of antipodes, and the form and stability of the earth. As Science put these things in their true light, Scripture, which had all along been interpreted in conformity with the current prepossessions, appeared to be in fault: till a closer examination into the real meaning of its language relieved it of the false interpretation which had been imposed upon it, and the harmony between Scripture and Science, although for a time they had appeared to be irreconcilable, was fully re-established.

The second class of examples in its character very much resembles the first, but belongs to a more recent period of discovery. Long-standing notions regarding some of the circumstances of the creation having been cleared away by the discoveries of Science, and Scripture being still fettered with the old interpretations imposed upon it in the days of ignorance, the cry of antagonism between Scripture and Science was again raised, and perhaps louder than ever. But in these instances also, the difficulty has been removed: and not only has Scripture been relieved of false interpretations, as in the first class of examples, but much light has been thrown upon its language and allusions, which would never have appeared but for these scientific discoveries.

Under the third class, I bring forward Examples in which Science, for a time, has in the hands of the self-confident made a retrograde movement. Conclusions have been put forth regarding the descent of all men from one blood, the differences of races since the flood, the original unity of language, the age of the human race, the superficial extent of the six days' creation, the origin of species, the origin of man, the origin of life, the uniformity of nature, design as indicative of creation, and certain numerical statements in the Pentateuch, which are contradictory to Scripture; and thus Scripture and Science were again declared to be at variance, till Science, under the guidance of wiser men, has corrected itself, and no want of harmony has been established.

CHAPTER I.

EXAMPLES FROM THE EARLIER HISTORY OF SCIENTIFIC DISCOVERY, IN WHICH SCRIPTURE HAS BEEN RELIEVED OF FALSE INTERPRETATIONS, AND THE HARMONY OF SCRIPTURE AND SCIENCE THEREBY RE-ESTABLISHED.

1. THE earliest instance of this kind which I shall produce affords a remarkable example of false notions of the celestial mechanism being incorporated in mistranslations of Scripture, in such a way as to consecrate error, and to sow the seeds of future perplexity by bringing God's two books into seeming collision.

<small>The Firmament.</small>

It is well known that the ancients conceived the heavens to be an enormous vault of transparent solid matter, whirling around the earth in diurnal revolution, and carrying with it the stars, supposed to be fixed in its substance. In accordance with this view, the Scripture was made by the LXX. to call the heavens στερέωμα (*stereoma*),—that is, something solid; and the Vulgate calls them *firmamentum*, which signifies the same. Josephus, in his 'Antiquities,' (professedly gathering his ideas from Scripture), in describing the creation calls the heavens κρύσταλλον (*krustallon*), *i.e.*

a sort of crystalline case.* Thus all seems to be in accordance, and Scripture and Science appear to agree and illustrate each other; till the light of later times pours in its beams, and, showing that space is not a solid mass, detects a seeming contradiction between the Word and Works of God. How is this to be met? Which is to yield? The popular solution, current to the present day, is this,—that Moses wrote, in matters of this description, not merely according to the appearance of things (which is true, and is the style which the most enlightened Science now uses in such a case), but in accommodation to the notions and prepossessions of the times. But is this the fact? Could not the Omniscient have put a correct word into the mind of His servant, as readily as one contradictory to fact? Let us turn to the word which the Holy Ghost has used by the pen of the inspired writer, and what do we find? that the original by no means implies, of necessity, a solid mass, but an EXPANSE :†—'And

---

* The following is from Josephus : 'After this, on the second day, he placed the heaven over the whole world, and separated it from the other parts; and he determined it should stand by itself. He also placed a crystalline [firmament] round it (κρύσταλλόν τε περιπήξας αὐτῷ), and put it together in a manner suited to the earth.'—*Joseph. Antiq.* lib. I. cap. i. § 1.

† The following is Pool's comment, and Gesenius' meaning of the word is given below :—

'6. *Fiat firmamentum.*]—Alii non firmamentum vertunt, sed expansionem, rem expansam seu extensam, eo modo quo aulæa expanduntur, ut tentorium quod funibus sustinetur ne decidat, vel sicut argentum malleo diducitur et attenuatur. Inde Deus dicitur *extendere cœlos.* Isa. xl. 22 et xlii. 5, et Ps. civ. 2. Grot. reddit τάσις (quæ vox Platonis est). רָקַע, est *expandere*. Laminæ expansæ appellantur רִקֻּעֵי פַחִים. Num. xvi. 38. *Expansum firmamentum* vertit A[insworthus]. Expansio hæc est diffusum corpus acris. Nam quid, nisi

God said, Let there be an expanse in the midst of the waters, and let it divide the waters from the waters. And God made the expanse. . . . And God called the expanse Heaven.' (Gen. i. 6-8.) So that, in fact, the inspired writer used the best possible word to express the actual appearance and state of things; but man, in his undiscriminating ignorance

aer, dividit aquas inferiores, *i. e.* mare, a superioribus? Nec aliud aeri nomen est Hebræis quam רָקִיעַ et שָׁמַיִם. Hoc nomen aeri tribui testantur Chald. par. in Ps. xix. et K(imchi) in Ps. lxxvii. *Quid mirabilius aquis in cœlo stantibus?* ait Plinius, l. 31. *Aves cœli* vocantur, Jer. vii., Os. ii., Matt. xviii. et xiii. Alii exponunt *firmamentum*, et accipiunt de orbibus cœlestibus. Complectitur tamen hæc vox etiam aerem vicinum, à cœlo in terram expansum, et suo loco quasi firmatum. ὁ vertunt στερέωμα, vel quia רָקִיעַ est στερεῶ, *i. e.* firmo, stabilio; ita vertunt ὁ Ps. cxxxvi. 6, Isa. xlii. 5, et xliv. 24; vel quia cœlum sæpe tentorio confertur, quod dicitur πήγνυσθαι (*i. e.* funibus ad paxillos in terram depactos firmari) quatenus expanditur Esa. xlvii. 5; vel potius a Syriaco usu רָקַע quod significat πιέζειν, comprimere, Luc. vi. 38. Et forte רָקַע Heb. primo significat *comprimere*, indeque *extendere*, nam premendo res extenduntur, ut laminæ æris.'— *Vide Poli Synopsis,* Gen. i. 6.

The extract below from Leo's translation of Gesenius' Hebrew Lexicon will give his idea of the meaning of the word. In his *comment*, in the latter part of this extract, Gesenius appears to side with the popular notion I have alluded to in the text; but with this we have nothing to do, but only with the meaning of the word, which he shows will well convey the idea of *expanse*, in the sense of open space or expanded atmosphere. Luther's translation, it will be seen, is the only one which does not convey a false idea, except our authorised version in the margin.

' רָקִיעַ m., more fully רְקִיעַ הַשָּׁמַיִם Gen. i. 14, 15, 17, *that which is distended, expanded* (from רָקַע) *the expanse of heaven; i. e.* the arch or vault *of heaven,* which, as to mere sense, appears to rest on the earth as a hollow hemisphere. The Hebrews seem to have considered it as transparent, like a crystal or sapphire (Ezek. i. 22; Dan. xii. 3; Exod. xxiv. 10; Rev. iv. 6); hence, different from the brazen and iron heaven of the Homeric mythology. Over this arch they supposed were the waters of heaven (Gen. i. 7, vii. 11; Ps. civ. 3,

of nature, has, by his successive versions of the Word of God, thrown a cloak of sacredness around his own error, in a way calculated to bring discredit upon the Holy Scriptures, as the discoveries of Science clear away the mists. Here, then, Scripture was right from the beginning, and all the confusion has arisen solely from human ignorance and misconception.*

---

cxlviii. 4). LXX. στερέωμα. Vulg. *firmamentum.* Luther, *Veste.*'— See Leo's *translation of Gesenius' Lexicon.* In *Aids to Faith* (pp. 220-230), Dr. M'Caul has brought his Hebrew learning to bear with effect upon this subject.

'Mr. Goodwin [in *Essays and Reviews*] wishes to fasten on the Hebrew word the sense of a "*solid* vault," as that sense which was always received until astronomy and modern geology taught men science; and he alleges that to translate it by the word "expanse', is a mere afterthought of the theologians [although it has been in the margin of our Bibles before modern astronomy and geology were thought of]! He says (p. 220), "It has been pretended that the word *rakia* may be translated *expanse,* so as merely to mean 'empty space.' The context sufficiently rebuts this." (!) Now what is the fact? The first translation of the Hebrew Bible made in modern days was that of Pagninus, who lived 400 years ago, and was one of the profoundest Hebraists of his own or any age. He translates this word *expansionem* in every instance. In the next century that extraordinary Oriental scholar (as ignorant of geology as geologists can possibly be of Hebrew), Arias Benedict Montanus, who had been appointed to revise the work of Pagnin for the King of Spain, again insisted on *expansionem* as the true meaning of this word *rakia.*'— *Literary Churchman,* April 1, 1861, p. 129.

* It is interesting to observe that the New Testament writers, who often quoted the Septuagint version *verbatim et literatim,* have been preserved from using this erroneous term στερέωμα to describe the celestial firmament; although it occurs in several places in that version of the Old Testament, and the New Testament writers had not scientific knowledge to avoid the error of themselves. The word is once used by St. Paul, but in an entirely different sense (Col. ii. 5), τὸ στερέωμα τῆς εἰς Χριστὸν πίστεως ὑμῶν,—'The steadfastness (or, as Alford renders it, the *solid basis*) of your faith in Christ.'

It has been suggested, that the fact that Moses (Gen. i. 21)

2. Another instance of the Scriptures having been drawn into this unworthy collision with the facts of Nature, is seen in the denial of the existence of Antipodes on the opposite side of the earth. I am not aware of any particular texts, unless one soon to be mentioned is excepted, which have been quoted to support this view; but no less a writer than the great Augustine, who in so many places* shows the greatness of his mind in not suffering Scripture and Nature to come into conflict, unfortunately brings the *silence* of Scripture to bear upon this question. He says that 'the story of there being antipodes, or men on the opposite side of the

Antipodes.

particularly specifies 'whales' among the creatures of the deep, indicates, also, that he wrote by inspiration, and was overruled to use language the minute correctness of which Science could alone illustrate; as this term might be taken as the generic representative of that remarkable class of sea-animals which are warm-blooded and suckle their young. This suggestion, however, cannot be sustained; for the Hebrew word תַּנִּין is used in the Old Testament in other senses; *e.g. serpents* (Exod. vii. 9, 10, 12); and, very frequently, *dragons*, described as living in ruined cities and desolate places, and, no doubt, meaning serpents in those places also, but by no means whales or sea-mammalia. It is a remarkable fact, however, that, while the word may be translated 'whales,' creatures of the cetaceous genus are at present found only in the upper strata of the earth. See this noticed in *Christian Observer*, May, 1867, p. 333, and take it in connexion with the view which I advocate further on, that the creation of the six days was the creation and preparation of the present order of things only.

* The following are specimens :—

'Si manifestissimæ certæque rationi velut Scripturarum Sanctarum objicitur autoritas, non intelligit qui hoc facit; et non Scripturarum illarum sensum (ad quem penetrare non potuit) sed suum potius objicit veritati; nec id quod in eis, sed quod in scipso velut pro eis invenit, opponit.'—*Aug. Epist.* 143, alias 7, *ad Marcellinum.*

'Respondendum est hominibus qui libris nostræ salutis calumni-

earth, where the sun rises when it sets to us, planting their footsteps opposite to our feet, is on no account to be believed:' and that 'even if the earth be a globe,' (a thing in his mind very doubtful), 'it does not follow that the opposite side is not an ocean; and, even should it be bare of water, it is not necessary that it has inhabitants; since the Scripture is in no way false, but secures belief in its narrative of the past, inasmuch as its predictions of the future are accomplished. And it is utterly absurd,' he adds, 'to suppose that any men should have crossed the vast ocean from this side to that, to establish the human race there as well as here.'* He appears to conceive,

ari affectant, ut quicquid ipsi de natura rerum veracibus documentis demonstrare potuerint, ostendamus nostris literis non esse contrarium; quicquid autem de quibuslibet suis voluminibus his nostris literis id est catholicæ fidei contrarium protulerint, aut aliqua etiam facultate ostendamus aut nulla dubitatione credamus esse falsissimum: atque ita teneamus Mediatorem nostram in quo sunt omnes thesauri sapientiæ atque scientiæ absconditi, ut neque falsæ philosophiæ loquacitate seducamur, neque falsæ religionis superstitione terreamur.'—*Aug. de Genesi ad literam*, lib. I. cap. xxi. § 41.

'Nunc autem servata semper moderatione piæ gravitatis, nihil credere de re obscura temere debemus, ne forte quod postea veritas patefecerit, quamvis libris sanctis sive testamenti veteris sive novi nullo modo esse possit adversum, tamen propter amorem nostri erroris oderimus.'—*Aug. de Gen. ad lit.* lib. II. *in fine.*

* '*An inferiorem partem terræ, quæ nostræ habitationi contraria est, antipodas habere credendum sit.*

'Quod vero et antipodas esse fabulantur, id est homines à contraria parte terræ, ubi sol oritur, quando occidit nobis, adversa pedibus nostris calcare vestigia, nulla ratione credendum est. Neque hoc ulla historica cognitione didicisse se affirmant, sed quasi ratiocinando conjectant, eo quod intra convexa cœli terra suspensa sit, eundemque focum mundus habeat et infimum et medium; et ex hoc opinantur alteram terræ partem, quæ infra est, habitatione hominum carere non posse. Nec attendunt, etiamsi figura conglobata et

that as Scripture does not tell us of any people on the opposite side of the globe, and he did not imagine that any could have traversed the boundless ocean, it must be concluded that there are no people there. But geographical research has divested this argument of all its force. In Behring's Strait a narrow sea exists, across which many an adventurous bark may have found its way even in the days of only primitive seamanship, and carried across to the furthest regions descendants of the sons of Noah, who spread forth on all sides to people the earth. Meanwhile Scripture, although it speaks of no nations but such as took their rise and dwelt on this side of the globe, presents no contradiction to the fact which actual observation teaches; for it nowhere says, that none ever had reached or ever would reach those furthest and then unknown regions.

A little more than a century later, Cosmas, monk of Alexandria, and a celebrated traveller, reasoned

rotunda mundus esse credatur sive aliqua ratione monstretur, non tamen esse consequens, ut etiam ex illa parte ab aquarum congerie nuda sit terra. Deinde etiamsi nuda sit, neque hoc statim necesse est, ut homines habeat : quoniam nulla modo Scriptura ista mentitur, quæ narratis, præteritis facit fidem, eo quod ejus prædicta complentur. Nimisque absurdum est ut dicatur, aliquos homines ex hac in illam partem Oceani immensitate trajecta navigare ac pervenire potuisse, ut etiam illic ex uno illo primo homine genus institueretur humanum. Quapropter inter illos tunc hominum populos, qui per septuaginta duas gentes et totidem linguas colliguntur fuisse divisi, quæramus, si possumus invenire illam in terris peregrinantem civitatem Dei, quæ usque ad diluvium arcamque perducta est, atque in filiis Noe per eorum benedictiones perseverasse monstratur, maxime in maximo, qui est appellatus Sem : quandoquidem Iaphet ita benedictus est, ut in ejusdem fratris sui domibus habitaret.'—*Aug. de Civitate Dei*, lib. XVI. cap. ix.

against there being antipodes from Gen. ii. 5. 'These are the generations of the heavens and the earth.' He said that this language is intended to comprise everything that is contained in the heavens and the earth; and that if there be antipodes, the heavens must contain the earth, and the sacred writer would have said simply, These are the generations of the heavens: with such slender arguments were men satisfied before the dawn of the days of exact science.*

3. Closely allied to this is the question already alluded to, which also exercised the ingenuity of the ancients, Whether the Earth is a Globe, or a vast extended Plane? or, which amounts to the same, Whether the heavens are a sphere surrounding the earth, or a wide-spread canopy overshadowing its extended surface? And there were not wanting advocates who appealed to Scripture to decide the question. There could be no doubt, for instance, when the Psalmist thus spoke of the Creator: 'who stretchest out the heavens LIKE A CURTAIN' (Ps. civ. 2). To these Augustine alludes, although he himself repudiates the appeal. 'It is commonly asked,' he says, 'of what form and figure we may believe heaven to be according to the Scripture. For many contend much about these matters, which with greater prudence our authors [meaning the sacred penmen]

Form of the Earth.

---

* The words of Cosmas in the Latin translation of his work are,—'Ait, "Hic est liber generationis cœli et terræ," quasi omnia iis contineantur, et universa qua in eis sunt cum illis significentur. Num si secundum fucatos illos Christianos cœlum tantummodo universa contineat, terram cum cœlo non nominasset, sed dixisset, "Hic est liber generationis cœli."'—See Lecky's History of *Rise of Rationalism*. Third edition. Vol. I. p. 296.

have forborne to speak of.' 'What is it to me,' he adds, whether heaven, as a sphere, on all sides environs the earth, balanced in the middle of the world, or whether, like a dish, it only covers and overshadows the same?' And he then throws out a salutary caution against appealing to Scripture in such cases, lest, misunderstanding the divine expressions, we should give interpretations, in physical subjects, which may prove to be contrary to fact, and so tempt others to suspect the truth of the sacred writers in more profitable matters.*

* 'Quæri etiam solet, quæ forma et figura cœli esse credenda sit secundum scripturas nostras. Multi enim multum disputant de iis rebus, quas majore prudentia nostri auctores omiserunt, ad beatam vitam non profuturas discentibus, et occupantes (quod pejus est multum) preciosa et rebus salubribus impendenda temporum spatia. Quid enim ad me pertinet, utrum cœlum sicut sphæra undique concludat terram in media mundi mole libratam, an eam ex una parte desuper velut discus operiat? Sed quia de fide agitur scripturarum, propter illam causam quam non semel commemoravi—ne quisquam eloquia divina non intelligens, cum de his rebus tale aliquid vel invenerit in libris nostris vel ex illis audierit, quod perceptis à se rationibus adversari videatur, nullo modo eis cætera utilia monentibus, vel narrantibus, vel pronunciantibus, credat—breviter dicendum est, de figura cœli hoc scisse auctores nostros quod veritas habet sed Spiritum Dei qui per ipsos loquebatur noluisse ista docere homines nulli saluti profutura.'—*Aug. de Genesi ad Lit.* lib. II. cap. ix. § 20.

The following remarks of Lactantius, a Christian writer at the end of the third and the beginning of the fourth century (or some writer using his name), against the rotundity of the earth and the existence of antipodes, afford a curious specimen of the arguments which sway the mind when devoid of what Dr. Whewell so aptly designates, in his 'History of the Inductive Sciences,' 'the appropriate idea :'—

'Quid illi, qui esse contrarios vestigiis nostris Antipodas putant, num aliquid loquuntur? aut est quisquam tam ineptus qui credat esse homines, quorum vestigia sint superiora quam capita? aut ibi quæ apud nos jacent, inversa pondere? fruges et arbores deorsum

4. The great controversy in which Galileo bore so conspicuous a part, regarding the motion of the earth, furnishes a further and very striking illustration, from the history of the past, of the mischief of bringing Scripture to bear upon scientific questions. We may be inclined, perhaps, to smile at the doubts and difficulties which beset men in those days on points which appear so simple to us, and which every child knows. But we must remember that they were good and learned men who debated on these matters. In the struggles of that period,

Motion of the Earth.

versus crescere ? pluvias et nives, et grandinem sursum versus cadere in terram ? Et miratur aliquis hortos pensiles inter septem mira narrari, quum philosophi et agros, et maria, et urbes, et montes pensiles faciant ? Hujus quoque erroris aperienda nobis origo est. Nam semper eodem modo falluntur. Quum enim falsum aliquid in principio sumserint, verisimilitudine inducti, necesse est eos in ea, quæ consequuntur, incurrere. Sic incidunt in multa ridicula : quia necesse est falsa esse, quæ rebus falsis congruunt. Quum autem primis habuerint fidem, qualia sunt ea, quæ sequuntur, non circumspiciunt, sed defendunt omni modo ; quum debeant prima illa utrumne vera sint an falsa ex consequentibus judicare. Quæ igitur illos ad Antipodas ratio perduxit ? Videbant siderum cursus in occasum meantium, solem atque lunam in eandem partem semper occidere, et oriri semper ab eadem. Quum autem non perspicerent, quæ machinatio cursus eorum temperaret, nec quomodo ab occasu ad orientem remearent, cœlum autem ipsum in omnes partes putarent esse devexum, quod sic videri propter immensam latitudinem necesse est ; existimaverunt rotundum esse mundum, sicut pilam et ex motu siderum opinati sunt cœlum volvi ; sic astra solemque, quum occiderint, volubilitate ipsa mundi ad ortum referri. Itaque et æreos orbes fabricati sunt quasi ad figuram mundi, eosque cælarunt portentosis quibusdam simulacris, quæ astra esse dicerent. Hanc igitur cœli rotunditatem illud sequebatur, ut terra in medio sinu ejus esset inclusa. Quod si ita esset, etiam ipsam terram globo similem ; neque enim fieri posset, ut non esset rotundum quod rotundo conclusum teneretur. Si autem rotunda etiam terra esset, necesse esse, ut in omnes cœli partes eandem faciem gerat, id est, montes erigat, compos tendat, maria consternat.

between reason and observation on the one hand, and Scripture, or rather Scripture falsely interpreted, on the other, and in the old prepossessions which the men of those days had to abandon, we see the very same causes at work, which still, under new circumstances, agitate and confuse religious but uninstructed minds. Nothing could be more clear, they then thought, than the testimony of Scripture—'the world also is established, that it CANNOT BE MOVED' (Ps. xciii. 1). Even so late an author as Calvin, the erudite and sagacious commentator, drew from this passage the inference that the earth is motionless.* The old Ptolemaic system, which had so blinded men for ages, chiefly under the authority of Aristotle, was only beginning about that time to meet its death-blow; and the new ideas had not yet reached the study of the learned reformer. Eleven centuries before him, when Pythagorean notions

Quod si esset, etiam sequebatur illud extremum, ut nulla sit pars terræ, quæ non ab hominibus ceterisque animalibus incolatur. Sic pendulos istos Antipodas cœli rotunditas adinvenit. Quod si quæras ab iis, qui hæc portenta defendunt, quomodo non cadunt omnia in inferiorem illam cœli partem? respondent hanc rerum esse naturam, ut pondera in medium ferantur, et ad medium connexa sint omnia, sicut radios videmus in rota; quæ autem levia sunt, ut nebula, fumus, ignis, a medio deferantur, ut cœlum petant. Quid dicam de iis nescio, qui, quum semel aberraverint, constanter in stultitia perseverant, et vanis vana defendunt, nisi quod eos interdum puto aut joci causa philosophari, aut prudentes et scios mendacia defendenda suscipere, quasi ut ingenia sua in malis rebus exerceant vel ostenteut. At ego multis argumentis probare possem nullo modo fieri posse, ut cœlum terra sit inferius, nisi et liber jam concludendus esset, et adhuc aliqua restarent, quæ magis sunt præsenti operi necessaria: et quoniam singulorum errores percurrere non est unius libri opus, satis sit pauca enumerasse, ex quibus possit, qualia sint cetera, intelligi.'—*Lactantii Omnia Opera*, *Oxon.* 1684. *Institut.* lib. III. cap. xxiv.

* *Ps.* xciii. 1.—'The Psalmist proves that God will not neglect or

had not been so entirely eclipsed, Augustine refers to the controversy thus : 'Some ask the question, Is the heaven at rest, or does it move ? If it moves, they say, how is it a firmament ? If it is at rest, how do the stars, which are supposed to be fixed in it, move from the east to the west?'* Augustine avoids coming to a decision, on the plea of want of leisure to discuss it, and absence of profit to his hearers. Mixed up, however, as the question is in the above statement of the case with the error regarding the firmament, it is doubtful whether they could have come to a correct result ; and we see the mischief which is likely to ensue from our taking our ideas of natural phenomena from Scripture language in the first instance, and shutting our eyes to the just conclusions of reason.

Other Scripture texts were forced into this unholy warfare. 'God . . . who laid the foundations of the

abandon the world, from the fact that He created it. A simple survey of the world should of itself suffice to attest a Divine Providence. The heavens revolve daily, and, immense as is their fabric, and inconceivable the rapidity of their revolutions, we experience no concussion—no disturbance in the harmony of their motion. The sun, though varying its course every diurnal revolution, returns annually to the same point. The planets, in all their wanderings, maintain their respective positions. How could the earth hang suspended in the air were it not upheld by God's hand ? By what means could it maintain itself unmoved, while the heavens above are in constant rapid motion, did not its Divine Maker fix and establish it? Accordingly the particle אף, denoting emphasis, is introduced—"Yea, he hath established it."'—*Commentary on the Psalms, Calvin Translation, Society's Edition.*

* 'De motu etiam cœli nonnulli fratres questionem movent, utrum stet an moveatur: quia si movetur, inquiunt, quomodo firmamentum est? Si autem stat, quomodo sidera, quæ in illo fixa creduntur, ab oriente usque ad occidentem circumeunt?'—*Aug. de Gen.* lib. II. cap. x.

earth, that it should NOT BE REMOVED FOR EVER,' (Ps. civ. 5). 'One generation passeth away, and another generation cometh; but the EARTH ABIDETH FOR EVER,' (Eccles. i. 4). Then the following were adduced to establish the correlative truth, as they supposed, that the sun is not at rest:—'In them hath he set a tabernacle for the sun, which is as a bridegroom COMING OUT of his chamber, and rejoiceth as a strong man to RUN a race. 'His GOING FORTH is from the end of the heaven, and his CIRCUIT unto the end of it' (Ps. xix. 4-6). 'The sun also ARISETH, and the sun GOETH DOWN, and HASTETH to his place where he AROSE' (Eccles. i. 5).*

The mischief which this appeal to Scripture did is incalculable. It sanctified error. It confirmed the mind in blunders regarding a fact in nature on which some of the ancients had clear and correct conceptions, till chiefly Aristotle and then Ptolemy also, even after a more complete theory had been suggested to him by

* They resorted to such arguments as the following curious piece of reasoning:—Hell, it had long been supposed, was in the centre of the world. Now, if the sun were at rest, with the earth revolving about it, then the centre of the world would be in the sun. So that hell would be in the sun, and therefore, in fact, be up in heaven —which was too absurd, they thought, to be believed. In laughing at such folly, let us beware lest we be guilty of the same *in our way*, notwithstanding all the light that knowledge gives us, and all the experience that the history of error and of well-intentioned but ill-directed zeal teaches us.

Another argument was, that heaven and earth are repeatedly mentioned in Scripture as correlative, like the centre and circumference of a circle. Thus 'the heaven and the earth' (Gen. i. 1), and in a multitude of other texts. Now, said they, the heavens, spread out as they are, must be the circumference; hence, the earth must be the *centre*, and therefore at rest.

Aristarchus of Samos, than had been to Aristotle, drew the veil of obscurity over the subject.* So that even Tycho, a name of eminence among philosophers in the days of Kepler and Galileo, was kept back from holding the true view, chiefly by his false estimate of Scripture language. †

All this conflict of ideas and opinions is now passed away; and Scripture stands unscathed in all its truth, simplicity, and beauty. All are agreed that its words require no apology, and call for no compromise. They speak intelligibly and correctly to learned and unlearned. Indeed, we may well pause to admire the wisdom with which Scripture phraseology has been chosen. Wisdom seen in choice of Scripture phraseology.

* Pythagoras, as stated by his follower, Philolaus, held that the earth is not motionless in the centre of the universe. His planetary theory was not, however, identical with the Copernican. He conceived that the sun, as well as the planets and the earth, revolved round a mass of fire in the centre of the system, invisible to us, because on the opposite side of the earth from the then inhabited part.

Aristarchus of Samos, in the third century, B.C., proposed a theory of the world exactly similar to the Copernican. This was subsequent to the time of Aristotle; but both Archimedes and Hipparchus rejected the theory of Aristarchus, as did also Ptolemy in the second century after Christ.—See Sir G. C. Lewis's *Historical Survey of the Ancients*, pp. 123, 124, 189, 252.

† See Sir David Brewster's *Martyrs of Science*. The freedom of Kepler's mind is nobly shown in the following words, quoted by Dr. Whewell:—'In Theology we balance authorities, in Philosophy we weigh reasons. A holy man was Lactantius, who denied that the earth was round; a holy man was Augustine, who granted the rotundity, but denied the antipodes; a holy thing to me is the Inquisition, which allows the smallness of the earth, but denies its motion; but more holy to me is Truth, and hence I prove, from philosophy, that the earth is round, and inhabited on every side, of small size, and in motion among the stars, and this I do with no disrespect to the Doctors.'

Human systems of religion have usually blended a false theology with some preposterous system of natural philosophy; and the application of true science is sufficient to explode the whole.* Now, not only has Scripture abstained from thus blending scientific teaching with divine, but, wherever it has incidentally touched upon the phenomena of the natural world, it has avoided the use of scientific terms, and has adopted a phraseology intelligible to all men in all ages. It

---

* 'Examine all the false theologies of the ancients and moderns; read in Homer or in Hesiod the religious codes of the Greeks; study those of the Buddhists, those of the Brahmins, those of the Mahommedans; you will not only find in these repulsive systems on the subject of the Godhead, but you will meet with the grossest errors on the material world. You will be revolted with their theology, no doubt; but their natural philosophy, and their astronomy also, ever allied to their religion, will be found to rest on the most absurd notions.'— *Theopneustia*, by M. Gaussen, chap. iv. sec. 6.

In the *Christian Observer* (May, 1870), the same thing is shown by pointing out that an uninspired writer in 2 Esdras, vi. 38–59, though taking his account of creation from Genesis, has not been able to restrain himself from making sundry additions of his own, which can all be proved to be false. He says (v. 40) God commanded light to come forth from His treasures, almost implying that light is a material substance, and not merely an effect; this is very different to the simple declaration, ' Let there be light, and there was light.' In v. 41 he says that God commanded the ' firmament to part asunder,' as if it were solid; whereas in Genesis it is said God ' made the firmament (expanse or atmosphere), and 'divided the waters' above and below it. He did not divide the firmament. In v. 42 he commits himself to saying that the sea is a seventh part of the land, whereas nothing of the kind is said in Genesis: and we know that the sea is about half as extensive again as the land. Indeed, the statement in Genesis, 'let the dry land appear,' seems almost to imply that the land was less extensive than the water out of which it emerged. The writer in the *Observer* in the same way compares Milton's description with Scripture, and shows how it departs from it, or adds to it, with scientific disadvantage.

speaks intelligibly simply because it speaks in such matters according to appearances.

When I say 'appearances,' I should explain this term, as it admits of two meanings, viz. (1) what *is seen*, (2) what *seems* to be. Thus I can conceive a person speaking as follows:—'When darkness came on, a comet *appeared;* and it *appeared* to be on fire.' The appearance in the first instance was a fact, something actually seen: the appearance in the second part of this description was only something which seemed to be, and was in reality in this case an illusion. It is obviously in the first of these senses that we use the word 'appearances,' when we say that it is according to appearances that Scripture speaks when it alludes to natural phenomena.*

The method of describing a phenomenon by appearances is as correct as any other method. There are two ways in which a phenomenon in nature may be described; either, first, with reference to the principles and laws of nature involved in the phenomenon: or, secondly, with reference to the facts or the results which an observer beholds. The first is called the scientific description; the second, the description according to appearances, or what is seen. These are equally real and equally true. The first is intelligible only to the scientific: the other can be understood by all in every age. This latter method, then, is the one

---

* In consequence of not seeing this distinction, Mr. Goodwin, in his *Mosaic Cosmogony* (p. 249), wrongly attributes to me the belief that 'appearances only, not facts, are described, in the Mosaic narrative.' There are several passages in the edition of this treatise which he used which might have saved him from falling into this mistake.

which Scripture adopts. It neither forestalls the knowledge, which it is left to man's reason and power of observation to acquire by long and patient investigation, and which, if used, would not become intelligible till the progress of science had made it so; nor does it adopt the language of current error and of false theory, in order to accommodate itself to the apprehensions of men. In matters of ordinary observation Scripture speaks the language of *sense*, not of theory: it uses the words of every-day life: it describes natural objects as they appear. It adopts the terms which the most scientific use in the ordinary intercourse of life, and not only so, but often even in their scientific writings, which would otherwise be encumbered and obscured with the most tiresome circumlocutions.* Here is no concession to vulgar prejudice, but an adoption of the usages of human language. What would be thought of even a scientific man, if he were to relate to his companions, that, on his looking towards a rain-cloud, he had beheld a beautiful phenomenon; that a succession of concentric circular arcs had sent forth to his eye, with inconceivable rapidity, an innumerable series of waves of various lengths, but so minute, that many thousands of them occupied only an inch, and that millions of them impinged upon his eye in every second of time? Whereas he might, in far easier language, and in language always intelligible all over

* Thus, I take down at random a volume of the *Astronomical Transactions*, and find a Paper by the Astronomer Royal, in which the following passage occurs:—' The meaning of the third term [in an astronomical formula under consideration] is, that the sun moves (independently of perturbations) in a small circle.'—*Roy. Ast. Society's Trans.* vol. x. p. 237, 1838.

the world, have simply adopted the popular, but no less true style, and said that he had seen a bow in the heavens, of colours varying from red to violet. The former would be a scientific way of describing the phenomenon, the causes and laws which were in operation to produce the result being introduced; and this would make it utterly unintelligible, till the undulatory theory of light was known, and even then it would remain so to millions, who would, nevertheless, have no difficulty in comprehending the other mode of description, which is quite as correct, though stating only what was actually seen. Although this illustration from the rainbow is somewhat far-fetched, it is not the less true and apt. It serves well to explain the difference between a scientific and a popular description of a phenomenon. The Copernican explanation of the solar system is now so entirely received, that no one for a moment doubts that the planets and the earth revolve about the sun. But this was not the case at one time, and such a statement of the phenomena of sun-rise and sun-set, as would require a knowledge of the physical law which causes the sun to be the centre, would then have appeared inexplicable, as the scientific description of the phenomenon of the rain-bow would be to most people now.

I take this opportunity of making some remarks about the terms *sun-rise* and *sun-set*, which I think are generally misunderstood. I believe that the sun does actually rise and set, go forth out of his chamber (viz. from below the horizon), and go down again. These terms have been so mixed up in past time, with the

<small>Sun-rise and sun-set correct terms.</small>

old and now exploded notion that the earth is immoveable, that most persons imagine that they do really imply that the sun moves and not the earth, and that they are now used only by accommodation and for convenience, and in fact are not true. This I do not believe. The terms are, I conceive, equally true, whether the Ptolemaic or the Copernican system be adopted. They are the description of the phenomena strictly according to what is seen, that is, according to appearances, and involve, as I hope now to show, no assumption whatever regarding the sun or the earth being the centre of the system. It is no doubt very difficult to divest the mind of a long-established persuasion, and to get it out of an old into a new train of thinking.

It is clear that the words *rise, set, fall, ascend, descend*, and suchlike, are RELATIVE terms, and do not by any means indicate of necessity ABSOLUTE motion in the body spoken of. In sun-rise the thing seen is the increase of distance between the top of the sun and the horizon; in sun-set the decrease of the distance, and the phenomena are described in terms of that object which most attracts attention, viz., the sun; and are called *sun*-rise and *sun*-set. This reference of them to the observer's *horizon*, a line fixed with regard to himself, involves no theory regarding the physical causes which produce the separation or approach; but is an obvious way of regarding the phenomena. The notion that the earth is immoveable has been so ingrained in former times in men's minds, and indeed to the present day so utterly imperceptible is the earth's motion in itself, that, as I have already said, these terms—*rise, set, fall, ascend, descend*—when applied to the sun, are generally supposed to imply that it does actually move in space, and that only for convenience and by accommodation the terms are still used. But, if our phraseology were to be reconstructed, or rather, if our present knowledge had been possessed when these terms were first introduced, I cannot think that others would have been used in preference. I cannot think that the position of the sun and other bodies would have been described by words, which would state that the standard of reference moved. For example, people would not have said, instead of 'See, the sun rises!' 'See, the horizon sinks!'—and simply for the reason mentioned, that the sun, and not the horizon, is the object which attracts attention. Indeed, we may feel sure of this, from the way in which similar language is perpetually used in cases, where no doubt ever existed as to which of the two bodies under consideration was fixed. In travelling towards a mountainous region, is it not the commonest of exclamations, 'Look, how the mountains rise!' they do not say '*seem* to rise,' but 'how they rise!' If one ship is pursuing another at sea, and the forward one is moving onward at a great pace, but the pursuer somewhat quicker, cannot we hear the captain of the hinder vessel saying, in sea language, 'How she rises!' although the ship pursued is going as fast as it can in the direction opposite to the *rise*, as the captain well knows. Suppose yourself sitting at the stern of a boat moving smoothly down a stream, and you call a friend at the other end to come to you, and he walks at the same rate at which the vessel moves down the stream, he is absolutely at rest relatively to the shore: and you are really moving down to him. When you call to him to come, you speak in relation to things around you, and it would be ridiculous for you to use other language, and to tell him to *stop* and to say that you were coming down to him, *meaning thereby* that he was to walk along the deck and you to sit still. So when we are sitting on the earth, and have our horizon, and are speaking of phenomena as affecting ourselves, there is nothing contrary to fact in saying that the sun rises from the horizon, omitting the additional and complementary fact, that the horizon is (like the vessel) smoothly sinking down at precisely the same rate, so as to leave the sun in space precisely where it was. The fact is, as I have said, all such terms as *sun-rise*, *sun-set*, *sun-standing-still*, express relative position, position relative to some standard, without any allusion whatever to physical causes and to motion caused by them. It would be a hazardous thing to trust, even in ordinary popular language, much more in Scripture phraseology, to scientific descriptions when reference is made to natural phenomena: for the theory may in the end prove to be false, or not sufficiently general to explain new phenomena, and will therefore require re-modelling and re-stating, whereas the description, recording appearances, or what is seen, will always stand, so long as men's senses remain the same. It so happens, that the theory, that the sun is the centre of the solar system now admits of indubitable proof. But it would, nevertheless, not be wise to surrender the general principle, that descriptions of natural phenomena, to be intelligible to all readers. should be couched in terms which involve no theory, however well established, but only in language which comes home to every ordinary observer. Such, as I have always received them, are the terms '*sun-rise*' and '*sun-set*,' in which there is no necessary reference to the absolute fixity of the sun or the earth; but only to the position of the sun relatively to the horizon, which is fixed with regard to the observer.*

---

* In a former edition, the following note appeared, which I should have omitted in the present one, as its object has been misconceived by Bishop Colenso, but that I wish to point out where his misconception lies:—

'Among many examples the language of the Sacred Historian in recording the miracle of Joshua is an excellent illustration of this [viz. of describing phenomena by their appearances]. *So the sun stood still in the midst of heaven, and hasted not to go down about a whole day.* . The accomplishment of this is supposed by some to have been by the arresting of the earth in its rotation. In what other words, then, could the miracle have been expressed? Should it have been said, "So the earth ceased to revolve, and made the sun appear to stand still in the midst of heaven?" This is not the language which we should use, even in these days of scientific light. Were so great a wonder again to appear, would even an astronomer, as he looked into the heavens, exclaim, 'The earth stands still!' Would he not be laughed at as a pedant? Whereas, to use the language of appearances, and thus to imitate the style of the Holy Scriptures themselves, would be most natural and intelligible. Conceive a vessel moving smoothly down a stream, and a man walking in a contrary direction on its deck at the same rate. What should we think of his asserting that he had never changed his situation at all? though this would be strictly and scientifically true. So a statement strictly scientific, in the case of Joshua's miracle, would have been unintelligible to common persons, and almost ridiculous in the ears of even the learned.'

My object in this note was, solely to adduce an *illustration of language*, but by no means to adopt any explanation of the mode in

There is another view of the wisdom shown in describing phenomena by their appearances only. Since the language of Science, even in its highest walks, admits of improvement, and has ofttimes called for correction as the field of discovery has widened, what epoch of knowledge should the Divine Author of Scripture have fixed upon as the best adapted for furnishing terms, if scientific phraseology were to be used? The more advanced the epoch, the longer would be the period through which Scripture would be

which the phenomenon was brought about. I admit that it looks as if I intended to adopt this explanation. Dr. Colenso thinks that I have given countenance to a 'view which every natural philosopher will know to be wholly untenable.' Though it is not my intention to advocate the explanation which supposes that the earth stood still, yet, that that event was impossible with Him who created the world I deny; nor would the 'profusion of miraculous interferences' which would have been necessary to counteract the *natural* effects of such an event (of which many will suggest themselves to a philosophical mind, not to say the melting of the whole mass with fervent heat, according to the mechanical theory of heat) be impossible or laborious to Him. But I hold no theory of explanation whatever of this miracle. Where a miracle is wrought the event is taken out of the pale of natural causes, and we are no more capable of reasoning upon it than we should have been able to foresee with what laws the world would be created, had we existed before its creation. Dr. Colenso considers Joshua's miracle to be one of the most prominent examples of Scripture and Science being at variance, and wonders why I do not enter upon it at large, but dismiss it so summarily. The answer is, that it was a miracle. I can no more ascertain how it was wrought than I can how our Lord walked upon the sea, or how His body ascended into the heavens.

Since the above was written (in 1864), I have been told that some, who are quite ready to admit that a miracle was wrought, stagger at the language used, viz. that the sun *stood still;* for, they say, the sun always stands still; what marvel, then, is there in this? My remarks in the text (about sun-rise, sun-set, sun-standing-still) sufficiently answer this.

unintelligible even to the learned, because it would anticipate human discovery. Moreover, were this the principle upon which Scripture was written, we should be in danger of finding our interest in the Sacred Volume DIVIDED between the truths which concern our moral state and eternal happiness, and the scientific mysteries hidden beneath these unintelligible terms. If, too, scientific phraseology were introduced into Scripture, reason would have no scope, or would be crushed at every turn. It was once the universal creed that the sun moved through the heavens. That it is absolutely fixed in space took its place. At the present day there is every reason to believe, from accurate astronomical observations, that the sun, with all its system of planets, is, after all, in motion. These are not conjectures, but the results of inquiry and reason. Whether the sun is absolutely fixed or not in space is, nevertheless, to this day unknown. We wait for science to give the answer. But if Scripture language is so chosen as to settle these questions at once, all such inquiries are hushed; the mind is cramped; reason justly feels her province invaded; and confusion follows. What admirable wisdom, then, is displayed by Him who knows the end from the beginning, who knows all laws, and foresees all their operations, since from Him they take their rise; in that He speaks to us of these things in terms always true and always intelligible!

I have dwelt at some length on this illustration from the Motion of the Earth. It is, however, highly instructive to fix our thoughts upon examples which the experience of the past furnishes; that we may

benefit by the mistakes of those who are gone before,
<small>This a highly instructive example.</small> learn wisdom in our own day, and see how we should behave in similar controversies which the march of discovery is perpetually stirring up amongst us. And no controversy is so well adapted for this purpose, as that regarding the motion of the earth. For no truth is at the present day more entirely and universally received; although no statement appears to be more contradictory to the letter of Scripture, and no physical fact is less palpable to the senses. There is, moreover, a difficulty involved in the belief of the earth's motion which only the mind habituated to scientific thought can thoroughly meet. If the earth revolves upon its axis in twenty-four hours, since its radius is 4000 miles, the equatorial parts of its surface must be moving at the amazing speed of 1000 miles an hour from west to east, and places in English latitudes at about 600 miles.* How is it, then, that the atmosphere rests quietly upon its surface, being subject only to local and occasional movements in winds and tempests, and those having no peculiar relation to the direction of east and west? How is it that our continents and oceans are not the scene of one incessant terrific tempest from the east, compared to which the most tremendous hurricane is but as the sighing of a summer breeze? The answer is, that the whole atmosphere itself in one mass is endowed, as well as the solid earth, with this prodigious velocity; and

---

* The velocity round the sun is still greater; more than sixty times that of the equator round the axis.

the winds and aerial currents, which we perceive, are but minor deviations from this average speed, occasioned by local and temporary causes. To the scientific, there is no difficulty in admitting this as one among many illustrations in nature of the primary laws of motion. To the unscientific, however, it is next to incomprehensible. How readily would the objectors in the days of Galileo have seized upon such an argument, as conclusive against the new-fangled errors, had they thought of it. But, notwithstanding this demand upon our belief that the atmosphere revolves at this prodigious speed, there is not one amongst us in these days who doubts for a moment that the earth revolves, and not the heavens. This is perhaps admitted by most persons under the pressure of the far greater demand which the other alternative would make upon them. For if the heavens revolve, and not the earth, we must believe that the stars move through millions of millions of miles within the twenty-four hours, even quicker than light itself: and also that their velocities, countless as these bodies are, are so adjusted to their distances, that they may preserve their several relative places, as seen from the earth, invariable from age to age. This shuts us up to the first alternative, that it must be the earth which revolves, and not the heavens. This is now the universal belief. It is received as the true view without hesitation, notwithstanding the difficulty of the atmosphere's revolving with such amazing velocity. Nor is the question regarded as an open one;· as one involving an unexplained difficulty, and therefore waiting for a better solution. The mind has been long habituated to the idea, and receives it. So marvellous is the effect of habit, even in thinking. 'Scientific views, *when familiar*, do not disturb the authority of Scripture,' however much they did upon their first announcement. 'Though the new opinion is resisted as something destructive of the credit of Scripture, and the reverence which is its due, yet, in fact, when the new interpretation has been generally established and incorporated with men's current thoughts, it ceases to disturb their views of the authority of Scripture, or of the truth of its teaching. .... And .... all cultivated persons look back with surprise at the mistake of those who thought that the essence of the revelation was involved in their own arbitrary version of some collateral circumstance in the revealed narrative.'* The lesson we learn from this example is this: How possible it is that, even while we are contending for truth, our minds may be enslaved to error by long cherished prepossessions. No man should act or believe contrary to his conscientious convictions. But it may sometimes be a great help to him to know, that it is possible he may be entirely in the wrong: and an example like this, regarding the Motion of the Earth, in which such strong views had been pertinaciously held on the side of error, but are now universally abandoned, is not without its use for this end.

* Dr. Whewell's *Philosophy of the Inductive Sciences.* Chapter on the ' Relation of Tradition to Palætiology.'

CHAPTER II.

EXAMPLES, FROM THE LATER HISTORY OF SCIENCE, IN WHICH SCRIPTURE HAS NOT ONLY BEEN RELIEVED OF FALSE INTERPRETATIONS, BUT HAS HAD NEW LIGHT REFLECTED UPON IT BY THE DISCOVERIES OF SCIENCE.

A COMPARISON of the discoveries of Geology with the statements of the first chapter of Genesis, has furnished within the present century several examples of apparent discrepancy between Scripture and Science, which further investigation has shown not to be real, while new light has been thrown by these discoveries upon the sacred text. Upon this subject I shall enter somewhat more at large than I have done in the early editions of this work, as the discussion has been revived by the publication of Mr. Goodwin's article on 'Mosaic Cosmogony,' in *Essays and Reviews*. Two questions have been mixed up together in this discussion : viz. Whether the teachings of the first chapter of Genesis are contradictory or not to the teachings of Science ; and, What is the undoubted meaning of the account in Genesis, interpreted scientifically ? It is with the former of these questions alone that the present treatise, according to its avowed object, is concerned. Although I do in the following pages express a strong preference for one of the two generally received modes of interpretation, it is sufficient for my purpose if I show that there is no real contradiction between Scripture and Science in the matter, whatever obscurity may still remain as to the precise meaning of the Scripture statement in a scientific point of view.

There are three leading particulars, in which the discoveries of modern Science are opposed to what were till lately the currently received views regarding creation, as gathered from this opening portion of the sacred volume.

1. The vast and unknown Antiquity of the Earth which geology has, undoubtedly, brought to light, Antiquity of the Earth. compared with which the 6000 years of its hitherto supposed existence are but as yesterday, is the first of those startling facts, which only a few years ago shocked many, who considered that such a conclusion was plainly repugnant to Holy Scripture.

2. The existence of Animals and Plants for many ages previously to the appearance of man on the Pre-Adamite animals and plants. earth, when first announced as a fact, was regarded as the fabrication of enemies of the sacred volume. The press teemed with attacks upon such reckless theorists; and crude hypotheses, hasty guesses, and ignorant assertions were thrust forward to take the place of facts. Every effort was made to crowd the countless tribes of creatures, which the rocks poured forth from their opened treasure-houses, within the six thousand years of man's existence; and to attribute their entombment to the

Deluge. But Science revolted at such summary work. Rushing waters were not the scene for deposits, in which all the bones and spines of the most delicate structures and the forms of leaves and plants in endless variety could be laid and kept unhurt. A deluge, and that, too, of only one hundred and fifty days' duration, was not the workshop in which strata ten miles thick could be formed and packed with their teeming population; it had neither time to do the work, nor room to hold the materials. Physiology, too, lent its aid. It was discovered that the buried species, at any rate below the higher (the tertiary) beds, differed essentially in their organization from the existing races. An order of things had then prevailed to which the present families could claim no relationship. Distinct acts of creative power must have called into life the existing beings, and those whose remains Science had brought to light. But Scripture records only one such epoch.

3. The existence of Light long prior to, not only the fourth day, on which we are told in Genesis the sun was made, but the first also, on which light was called forth, was another discovery which perplexed even philosophers, and which the multitude indignantly denied as repugnant to the simplest and plainest declarations of Holy Writ. Geologists found that the exhumed remains of animals, belonging to ages long gone by before man's appearance, had *eyes:* and it was argued that eyes were for use; that light was necessary, and that light must have existed. But all this seemed directly contrary to Scripture, which spoke thus of the

Existence of light before the Six Days' creation

first day. 'And God said, Let there be light, and there was light' (Gen. i. 3); and of the fourth day, 'And God made two great lights; the greater light to rule the day, and the lesser light to rule the night: (He made) the stars also. And God set them in the firmament of the heaven, to give light upon the earth, and to rule over the day, and over the night, and to divide the light from the darkness' (Gen. i. 16–18).

These are the chief difficulties which modern Science has advanced against the long-standing interpretation of the beginning of Genesis. There are two classes of interpreters who have endeavoured to remove the perplexity thus arising, and to show that Scripture and Science are not really at variance, when rightly interpreted. As Science reveals new phenomena, opens up new ideas, and makes new demands, not only do these require a searching scrutiny, but also the hitherto received interpretation of Scripture calls for re-examination. In this way, while it not unfrequently turns out that scientific men have been premature in their generalizations, it also sometimes happens that Scripture is seen to have been subjected to incorrect glosses, from which it is liberated by the discoveries of Science. The torches of nature and reason shed their light, in such instances, upon the letter of the Sacred Volume itself; and God's two books of Nature and Revelation, which appeared for a while to contradict each other, are found to be in harmony.

The first class of interpreters conceive that no violence is done to either the letter or the spirit of Scripture, and that all emergent difficulties are met,

Two theories of explanation.

PRE-ADAMITE CREATURES, AND THE SUN'S AGE. 45

by imagining that an interval of time of untold duration occurred between the first creation of all things 'in the beginning,' as announced in the first verse of Genesis, and the state of disorder into which the earth had fallen, as described in the second verse: and this view is much strengthened by the feeling, that the Almighty can hardly have created the earth in a state of confusion and chaos.* From this condition the Almighty raised the earth into one of beauty and order by the six days' work described in the subsequent verses, and so prepared it for the reception of His rational creature Man, whom he brought into existence and placed in the garden of Eden on the sixth day. This view was propounded by Dr. Chalmers half-a-century ago;† and was adopted by Dr. Buckland, Professor Sedgwick, and

First or Natural-day theory.

---

* *Tohu va bohu*, that is, *without form and void;* or, literally, 'desolation and emptiness.' Heb. For illustration of the meaning of this expression, see the description of the land of Israel as desolated and depopulated by Nebuchadnezzar, Jer. iv. 23–26. It is worthy of remark, that when the Almighty says, 'He created it (the earth) not in vain—He formed it to be inhabited,' (Isa. xlv. 18); the words are literally, He created it not *tohu*, but He formed it for habitation.

† Chalmers' Works, vol. xii. p. 369; vol. i. p. 228; Buckland's Bridgewater Treatise, chap. ii.; Sedgwick's Discourse on the Studies of the University of Cambridge. It has been pointed out to me that Dathe also gives this view in his translation of the Pentateuch (1791). His translation is, 'Principio creavit Deus coelum atque terram. Post hac vero terra facta erat vasta et deserta.' In a note he adds this comment, 'Non describitur prima telluris nostrae productio, sed altera, sive ejus restauratio.' I may add that the hypothesis, that there was a wide interval of time between the first and second verses, is not a modern one merely to meet the requirements of Science, but was suggested by some of the early Christian Fathers, as pointed out by Dr. Pusey in a note in Buckland's Bridgewater Treatise.

other eminent men, as an easy and sufficient solution of the difficulty. According to this hypothesis, it is supposed that the generations of animals and plants which are stored up in the earth's strata lived and perished in that interval of time of unknown duration which preceded the six days' creation, and that Scripture is altogether silent regarding them. The difficulties, therefore, which the first and second of the discoveries of geology, regarding the Age of the Earth, and the pre-existence of animals and plants long before Adam, gave rise to, are altogether removed. With reference to the third difficulty, arising from the discovery that Light also existed ages before Adam, and not only six days previously, it may be observed, in the first place, that it is not said that light was *created* or *made* at all, it is *called forth*, it is commanded to shine out of the darkness which was upon the face of the deep. 'Let there be light, and there was light;' or, Let light appear, and it appeared. Light is not a substance, but an effect. To suppose that the luminiferous ether, the vibrations of which cause light, was at this moment created, is not necessary; but simply, that the ether, which had been in a state of quiescence on the surface of the earth during the darkness, was caused to vibrate, so as to send forth light, as air vibrates, when properly acted on, and sends forth sound. Nor, with reference to the fourth day, is it said that the sun and moon and stars were *created* on that day: the word is '*made*'—'God MADE two great lights' (Gen. ii. 16)—the Hebrew word elsewhere signifying *appointed, constituted, set for a particular purpose or use;* and never once, in the one hundred and fifty places where it occurs

in the book of Genesis, is it used in the sense of
*created.**
This being premised, the account of the six days'
work may be paraphrased as below. The language of
this passage of Scripture consists of statements, and not

* 'There are three words employed in the Old Testament in
reference to the production of the world—*Bará*, he created; *Yatzár*,
he formed; *Asáh*, he made—between which there is this difference,
that the two last may be, and are, used of men. The first word, *Bará*,
is never predicated of any created being, angel, or man, but exclusively appropriated to God, and God alone is called *Boré*, בֹּרֵא,
Creator. Creation is, therefore, according to the Hebrew, a Divine
act—something that can be performed by God alone. In the next
place, though, according to its etymology, it does not necessarily
imply a creation out of nothing, it does signify the Divine production
of something *new*, something that did not exist before. See Num. xvi. 30;
Jer. xxxi. 22.'—Dr. McCaul, *Aids to Faith*, p. 203; also p. 212. This
remark that *bará* does not necessarily imply creation out of nothing is
made also by M. Max Müller in his *Chips from a German Workshop*, vol. i.
chap. vi.; from which it appears that the opinion of scholars has on
this subject changed, and that though it was once thought that *bará*
meant created out of nothing, this exclusive meaning is no longer
received. I may add, however, that the connexion in which the word
stands in Genesis shows that *in that place* it means 'created out of
nothing,' because 'in the beginning' means before anything existed
but the Divine Being.

The word occurs eight times in Genesis, and is in these places always
rendered by our translators *created*. It occurs forty times more in the Old
Testament; and in thirty-two of these it is rendered *created*; in three,
made; and in the other five it has various meanings. The second word
occurs three times in Genesis (ch. ii. 7, 8, 19), and is translated *formed*.
The third word, which occurs 154 times in Genesis, is not once rendered
created; it is eighty-eight times *did* or *done*; forty-five times *made*; and
twenty-one times has other meanings, regulated by the context. This
word occurs about 2700 times in the Old Testament, and I believe is not
once translated *created*. In short, the first word (*bará*), appears to be
used only when something *new* is made, which did not exist before;
the third word (*asáh*) appears to have a more general application, meaning
made, without specifying whether absolutely new or not: it seems never
to be translated *created*, so as to confine it to that idea; but it is,

of explanations; and, therefore, how far it pleased the Almighty to work by ordinary means, as in the daily government of the physical world at present, and how far by miracle, it is impossible to say. All I propose to

nevertheless, occasionally, used of things created, as in Gen. i. 26, compared with v. 27.

It is the third of the three words which is used in the Fourth Commandment, and not the first. *For in six days the Lord* MADE *heaven and earth, the sea, and all that in them is.* The work of the six days is here described, in which God did not now bring the world into existence, having done that 'in the beginning,' but re-moulded previously existing matter, and prepared the earth, and the sea, and the clouds, and the atmosphere (or heaven), for man's reception.

Dr. Colenso, at p. 92 of his Part IV., represents me as basing my statement, that Gen. i. 1 refers to the original creation of matter, 'in the beginning,' and the Fourth Commandment to a subsequent re-arrangement of that matter in the six days, on the *English Translation*. What I say is this, that in the passages, 'In the beginning God created (*bará*) the heaven and the earth,' and Exod. xx. 11, 'In six days the Lord made (*asáh*) heaven and earth,' not only are two different words—'created' and 'made'—used in our translation, but two different words are also used in the Hebrew; and that, by a collation of all the passages where these two words occur, it may be seen that the second (*asáh*) has a wide range of meanings, while the other (*bará*) has not. In this difference of words I find, I think, a *confirmation* of the thought, derived from other and independent premises, that in Gen. i. 1, the original creation of matter may be referred to, while in Exod. xx. 11, only a re-arrangement of the same matter may be spoken of. It might be inferred from Dr. Colenso's remarks on what he quotes from my Fourth Edition, that the fact of the sacred writer using these two distinct words is the ground of my argument. This is not the case. This philological basis is too slight to erect upon it such an inference: but it may be regarded, as far as it goes, as confirming the thought arrived at in another way. With reference to his remark in p. 94, I would say, that though this view does not satisfy him, it does me till a better is propounded. I do not feel it to be a 'broken reed .... piercing .... the hands that leant upon it.' Exod. xx. 11 might well be translated, 'For in six days the Lord fashioned heaven and earth, the sea, and all that in them is, and rested the seventh day.'

do is to show how it is possible to translate the language into that of physical occurrences, on the supposition that the days are natural days.

On the first day, while the earth was 'without form and void,' the result of a previous revolution on its surface, 'and darkness was upon the face of the deep,' God commanded light to shine upon the earth. This may have been effected * by such a clearing of the thick and loaded atmosphere as would allow the light of the sun to penetrate its mass with a suffused illumination, sufficient to dispel the total darkness which prevailed, but proceeding from a source not yet apparent on the earth. On the second day a separation took place in the thick vapoury mass which lay upon the earth, dense clouds were gathered up aloft, and separated from the waters and vapours below by an 'expanse,' the word rightly substituted in the margin of our Bibles for 'firmament.' On the third day, the lower vapours, or fogs and mists, which still concealed the earth, were condensed, and gathered, with the other waters on the earth, into seas—we may suppose by the upheaval of the ocean-bed in some places—and the dry land appeared. Then grass and herbs began to grow. On the fourth day, the clouds and vapours were so rolled into separate masses, or even altogether absorbed into the air, that the sun shone forth in all its brilliancy, the visible source of light and heat to the renovated earth,

---

* This view of the light of the sun, temporarily obscured, struggling through the earth's vapoury atmosphere, and not shining fully on the earth till the fourth day, is not peculiar to the natural-day interpreters.

while the moon and the stars gave light by night; and God appointed them henceforth for signs and for seasons, for days and for years, to the rational beings whom He was about to call into existence; as He afterwards set or appointed the rainbow, which had appeared ages before, to be a sign to Noah and his descendants. On the fifth and sixth days the waters and the earth brought forth living creatures, and man was created.

'We hold the week of the first chapter of Genesis,' writes Dr. Chalmers, 'to have been literally a week of miracles; the period of a great creative interposition, during which by so many successive evolutions the present economy was raised out of the wreck and materials of the one which had gone before.' That the Creation of Man was a signal act of Divine power, not brought about by secondary causes, we must, at any rate, admit. Why, then, should there be any hesitation in supposing that the great changes which preceded this act were due to an equally direct exercise of Creative Agency? It has been said, that if we have to accept the theory that the week of creation was a week of miracles, the whole question is placed beyond our criticisms; and we may just as readily at once admit, that all the tribes of creatures which geology reveals were created during those six natural days. But an examination of the fossils themselves must soon dispel such a notion. They exhibit the varied effects and accidents of ordinary vegetable and animal life to such a degree, as to make it far more difficult for us to suppose that these circumstances have been *imitated* in a miraculous existence and destruction of these creatures in a few days, than that the physical changes, as above described, may have taken place in the same time. That the fossils should have been created as they are found, with scattered bones and broken shells, is a notion, of course, too foolish to be entertained for a moment.

The particulars of the above paraphrase may be capable of improvement. Hereafter, some better one may be advanced upon this natural-day hypothesis. I have given this, as I have already said, only in order to indicate, that on the hypothesis of the days being natural days, a consistent interpretation can be given to the passage, as an account of actual occurrences, and that the account so interpreted advances nothing really contradictory to Science. I do not consider that Scripture receives any confirmation from Science by this interpretation; but I contend that it meets with no contradiction from it. According to this view, the first chapter of Genesis does not pretend (as has generally been assumed) to be a cosmogony, or an account of the original creation of the material universe. The only cosmogony which it contains, in that sense, is confined to the sublime declaration of the first verse: 'In the beginning God created the heaven and the earth.' Then stepping over an interval of indefinite ages, with which the human race has no direct concern, the inspired record proceeds at once to narrate the events preparatory to the introduction of man on the scene, employing phraseology strictly faithful to the appearances which would have met the eye of man, could he have been a spectator on the earth of what passed during those six days.*

The other method of interpretation has been adopted by various writers. According to this view, the six days <small>Second or period-day theory.</small> are imagined to be, not days of the ordinary kind, but periods of enormous duration, and not necessarily of equal length. In this way the first of the three difficulties I have mentioned is at once met; since, on this view, no limit whatever is assigned to the age of the earth. In support of this it is argued that the word 'day' is not always used in Scripture to mean a period of twenty-four hours, and that even within the space of the first thirty-five verses of Genesis the word has three separate meanings, viz. a day of daylight; a day of an evening and morning (Gen. i. 5); and a day which seemingly includes the whole six days of creation (Gen. ii. 4). The six days' narrative is taken to be an account of the creation, during these enormous periods of time, of the innumerable animals and plants which are found fossil in the strata; which are, therefore, not passed over in silence

* It has been suggested that even v. 1 may refer to the six days' creation, and that there is no interval between that and the second verse. In this case geological time would be prior to v. 1. This interpretation would have the advantage of making the language of the fourth commandment coincide more simply with that of the first chapter of Genesis, than when we have to exclude the first verse, and point out that the word 'made' is used, and not 'created.' But it is obvious, I should say, that St. John, in using the phrase 'In the beginning' at the commencement of his Gospel, made a distinct reference to the same phrase at the commencement of Genesis. In St. John's Gospel, there can be no question that it means in the beginning of all created things, and not the present order of things, as the declaration of the eternity of the Son of God is the apostle's object. This, then, must be the meaning in Genesis.

in this account, as they are supposed to be according to the other hypothesis. The second of the three discoveries of geology is thus accounted for. Mr. Miller, one of the advocates of this theory, considers that he can identify the work of the third, fifth, and sixth days respectively—when plants (i. 11-13), sea-monsters and creeping things (i. 20-23), and cattle with beasts of the earth (i. 24, 25) were made—with what geologists call the palæozoic, the mesozoic, and the kainozoic periods,* in which the copious flora producing the coal-measures, the huge saurian animals, and mammalia, are the distinguishing fossils. By 'evening and morning' he understands a diminution and increase in the existences, vegetable or animal, by which the periods are most prominently characterised. No meaning is assigned to these terms for the other three days—the first, second, and fourth. The third difficulty, regarding the non-appearance of the sun, moon, and stars till the fourth period, although light must have existed previously, is met very much as under the former method of interpretation, though without so simple an explanation of the cause of their obscuration throughout the long interval of the first, second, and third periods.†

Both these theories of interpretation, it will be seen, furnish a solution of the three difficulties I have enumerated, as advanced by modern Science against

---

* That is, three geological ages, in which the fossil creatures are supposed to be of ancient, middle, and recent types; corresponding with the Primary, Secondary, and Tertiary divisions of rocks and formations.

† *Testimony of the Rocks*, p. 152.

the old method of understanding this chapter. The charge against Scripture and Science that they are here at variance, falls therefore to the ground, as no contradiction is established. It is clear, however, that both these theories of explanation cannot be true, though neither may advance anything positively at variance with this brief statement of Scripture. Mr. Goodwin makes this difference of interpretation a matter of triumph, in his Essay on 'Mosaic Cosmogony,' and founds on it the illogical conclusion that both are wrong; and more than this, that the Mosaic account itself is untrue! He entirely passes by the several points in which the interpreters concur: viz. that the account in Genesis is true; that it was communicated to the writer by inspiration; that it teaches that matter is not eternal; that God created matter in the beginning; that the beginning may have been, and probably was, countless ages ago; that the document describes a creation which was distributed over six portions of time; that man was created out of the dust in the sixth period; that the Sabbath was instituted, for the benefit of man, in commemoration of this work. The points on which they differ are these: (1) whether the six days are ordinary days or not: (2) whether the brief account of what occurred in these six periods is sufficiently full to justify us in expecting to find in the records of terrestrial changes such corresponding traces of phenomena as may enable us to test the truth of the narrative. It is upon these points alone that the theories are 'mutually destructive.' But these are points which affect the explicitness of the narrative, not its truth. Both accounts agree in considering that the general idea of orderly succession was mainly what the narrative was intended to convey.

Though, however, I write in this general way when arguing against a common opponent, without any hesitation I adopt the first of the two methods of interpretation. I proceed now to give at length my reasons for this.

Before entering upon a comparison of these two methods of exposition.* I would observe, that it is difficult to see any good ground for sup- Genesis i. not a vision. posing, as some have done, that the revelation of these truths was made to Moses by vision. That they were supernaturally communicated is evident, as they refer to transactions which occurred before the creation of man; but one method of communication is as easy to the Almighty as another; and no special reason is apparent why that of vision should have been chosen. In other cases, where information was given in visions, this is expressly stated by the prophet to whom the vision was vouchsafed. But here there is not the slightest intimation to that effect. The verses in question bear all the marks of being a plain narrative, precisely such as an observer, had he been present, would have given. This mode of description is the one we often adopt in describing scenes to others who have not witnessed them: we treat them as actual spectators, at the time, of what we are describing. This method gives the liveliest and most durable impression of what we are describing. In

---

* Should my readers not be interested in this discussion, they may pass over 32 pages to page 86, where it ends.

the fourth commandment, these verses are, unquestionably, referred to as history.

Dr. Rorison* has suggested that they are a Psalm of Creation; and, after others who have preceded him in this, he points out that the verses follow the rules of parallelism, such as Bishop Lowth discovered in the Psalms and the Prophets. There can be no objection to this, if the psalm be taken to be an historical psalm. If I mistake not, however, Dr. Rorison does not take this view, and in so far I decidedly differ from him. He seems to consider that there are no chronological marks whatever in the narrative; that the expression, 'The evening was, and the morning was,' is merely a poetical refrain, to mark the change of chorus where the subjects of the poem change; and that the number seven is mystic.†

I will now enter upon my reasons for rejecting the period theory of explanation, and for accepting the natural-day theory.

Objections to the period theory.

The fact of the alternations of light and darkness being distinctly defined as Day and Night in Gen. i. 5, is, in my mind, a strong argument in favour of natural days, and against periods. For what is the primary meaning of Day and Night? —and surely the primary and currently used meaning must be designed here, where they are given as definitions. Any other meaning (such as in John, ix. 4: 'I must work .... while it is day, the night cometh wherein no man can work') is secondary and figurative, and derived from the primary one.

* *Replies to Essays and Reviews*, pp. 281-286.
† Ibid. pp. 333-336.

The use assigned to the sun and moon and stars, which are made to shine upon the earth on the fourth day, viz. to be 'for signs and for seasons, and for days and for years' (i. 14)—seems clearly intended for the advantage of God's intellectual creature, man. But upon the period system of interpretation, he was not created to enjoy this benefit for myriads of years after it was prepared for him—all the animal existences on the earth, the work of the fifth and sixth days, intervening before man's appearance. There is something incongruous in this.

The visible appearance of the sun on the fourth day, in the midst of the week, furnishes a conclusive argument for natural days, if the term 'and evening was, and morning was'—which occurs six times—is to be interpreted consistently throughout. For it may be fairly assumed that the heavenly bodies began at once to fulfil the functions assigned to them, viz. 'the greater light *to rule the day*, and the lesser light to rule the night' (i. 16); and therefore the fifth and sixth days, at least, must have been ordinary days of twenty-four hours each; and as they were of sufficient length for the works belonging to them, the first, second, third, and fourth may well have been sufficient for *their* respective works. The description, 'evening was, and morning was,' being the same, the day must be homogeneous.

Mr. Birks well remarks, that ordinal numbers never occur either in the Bible or elsewhere, when words of time are used as indefinite periods. And he adds, that the reason is plain : two, three, four indefinite periods make only one indefinite period. The simple fact that the days are numbered from the first to the sixth, is thus a clear proof that definite periods or days are meant.

The dominion given to Adam over the creatures, the creation of which had been described, and the use assigned to the fruit-trees, viz. that they should be for food to Adam (i. 28, 29), can, by no contrivance, be made to refer to the flora of the coal-measures, and the saurians and mammals which, according to the period-theory, had been for ages buried in the earth as fossil deposits!

The wording of the fourth commandment appears to me opposed to the view of periods. The spirit of the commandment, I admit, is the same whatever the duration of the days be. But the wording appears to forbid the period interpretation. Exod. xx. 8 : 'Remember the Sabbath-day to keep it holy 9. Six days shalt thou labour and do all thy work. 10. But the seventh day is the Sabbath of the Lord thy God: in it thou shalt not do any work, thou, nor thy son, nor thy daughter, thy man-servant, nor thy maid-servant, nor thy cattle, nor thy stranger that is within thy gates: 11. For in six days the Lord made heaven and earth, the sea, and all that in them is, and rested the seventh day: wherefore the Lord blessed the Sabbath-day, and hallowed it.' Is it not a harsh and forced interpretation to suppose that the 'six days' in v. 9 do not mean the same as the 'six days' in v. 11, but that in this last place they mean periods? In reading through the eleventh verse, it is extremely difficult to believe that the 'seventh day' is a long period and the 'Sabbath-day' an ordinary day; that is, that the same word 'day' should be used in two such totally different senses in the same short sentence, and without explanation.

There is one argument advanced for the period-theory which appears to me hardly worthy of being mentioned. It is said, that God's work of creation is not going on now, and that the formula 'and evening was, and morning was,' is not recorded for the seventh day, because His Sabbath rest is still going on. Hence the seventh day of the creation-week being a long period of time, the six working days must have been so too. But there are two replies to this. It proves too much. God's rest from the six days' work will last, through eternity; and therefore each of the six periods should have been of eternal duration to keep up the analogy. Again; it is the fact of a *cessation* after six days' work which is taught by the analogy of the creation-week, and not the *length* of the rest. Six days did God work, and then ceased. Six days we are commanded to do our ordinary work, and then to rest. The Almighty's work during the creation-week is once done and not repeated. Man has again to enter upon his ordinary occupations, and to spend his life in a cycle: and therefore the length of his rest is fixed, and it is appointed to be *one day* after six (Gen. ii. 3; Exod. xx. 10), when he begins his routine again. Too much is made, moreover, as it appears to me, of God's now resting from work, and human ideas are mixed up with divine in an unwarrantable manner. Surely the upholding of all things, the ordering of all things in heaven and earth, and the work of redeeming a ruined world, are undertakings as arduous — I speak with reverence, and am forced into this strain by the language of the opposite side—as the creation of the world, to Him who 'spake and it was done, who commanded and it stood fast' (Ps. xxxiii. 9). Mr. Miller, indeed, justly feeling the greatness of the scheme of redemption, is unable either to leave it out of the estimate of the Divine doings or to abstain from calling it a 'work.' His theory, therefore, of periods, which compels him to consider that God is now resting from His work of creation, compels him also to adopt the inconsistent course of assigning other work in the new creation now going on. 'The work of Redemption may be the work of God's Sabbath-day.'*

These six arguments which I have given against the period-theory are quite independent of any geological facts or speculations. The advocates of that theory draw their chief arguments from this source. Arguments can be drawn on the opposite side also from geological research. The following is an example.

The late M. D'Orbigny, in his *Prodrome de Palæontologie*,† by an elaborate examination of vast multitudes of fossils, gives reason to believe that there have been at least twenty-nine periods of animal and vegetable existence, that is, twenty-nine creations,

---

* *Testimony of the Rocks*, p. 153. See *Christian Observer*, July 1858, and *The Bible and Modern Thought*. Second edition.

Bishop Wordsworth, in his comment on Gen. i. 5, after observing that there must have been death in the geological periods, says, 'But the Days of Creation, as represented in the Book of Genesis, are not Days of Death, there is no place for death in them; they are days of creation only: and God saw everything that He had made in them, and behold it was *very good*, v. 31. But death is not good; it is evil: therefore, God saw no death in them; it was not there.'

† See Lardner's *Museum of Science and Art*, vol. xii., pre-Adamite Earth, pp. 5, 37, 38, 120, 155.

separated one from another by catastrophes which have swept away the species existing at the time, with a very few solitary exceptions, never exceeding one-and-a-half per cent of the whole number discovered, which have either survived the catastrophe, or have been erroneously designated. M. D'Orbigny states that both animals and planets appeared in every one of these twenty-nine periods. This is quite irreconcilable with the period theory. The parallel is destroyed both in the number of the periods—thirty, including the azoic, instead of six—and also in the character of the things created. To say that only the more important ones were announced in Genesis, is to resign the office of interpreter altogether. But I consider geology to be a science so young, still growing and so frequently shifting its ground, that arguments drawn from the speculations it gives rise to are not to be trusted in this grave inquiry, in which Scripture and Nature are brought into comparison.

Besides these objections to the period theory, it appears to me that the several methods of applying that theory all fail in some important particular. Thus, to begin with Mr. Miller's method. How ill do his three illustrations from the fossils of the palæozoic, secondary, and tertiary deposits, with their dawn and decline, suit the description of the work of the third, fifth, and sixth days, with their 'evening and morning.' On the third day grass and 'the fruit-tree yielding fruit' (v. 11), meant 'for meat' for man (v. 29), were created. Though edible fruit might have been found in the palmaceæ, cycadaceæ, and coniferæ, which

Each method of applying the period-theory fails somewhere.

abound in the coal measures, yet, what is to be made of the 'evening and morning?' If 'evening and morning' of these periods mean the decline and dawn of species (it is necessary to invert the terms from their natural order), what can this expression mean when used for the other periods, when no plants or animals were created? Assuredly, the Hebrews could derive no other idea from the words than that the literal evening and morning of the day were meant. There is another point, too, in which Mr. Miller's illustrations fail. They are by no means characteristic of the *whole* of the respective periods to which they belong, but only of some part, and that, perhaps, only a small part, and are obviously selected because they best suit the description in Genesis. And more than this; this interpretation would make it appear that the strata testify to the fact, that the earth was first covered with a flora, and that afterwards living creatures appeared (as described in Genesis); whereas geologists inform us that animal and vegetable life first appeared on the earth at one and the same time.

Let us next take Dr. M'Caul's interpretation.* I cannot but think that his accommodation of the nebular hypothesis of Laplace—which, though a highly ingenious hypothesis, is, after all, a mere hypothesis, with various scientific difficulties besetting it,—to the elucidation of the account of creation, is very hazardous. Is it wise to mix up Scripture interpretation with a theory of planetary evolution so utterly uncertain as

* *Aids to Faith*, p. 212.

this is? Our first-rate works on astronomy do not even allude to it.* Nor is the aid which the theory, were it true, would bring to his interpretation of any use. The theory is briefly this: that myriads of ages ago, the planetary space was filled with an enormous gaseous globe of great heat revolving on an axis; that as the heat radiated into space the mass contracted, and the velocity of rotation therefore increased; that during the process of contraction the globe threw off a ring of matter at its equator, when the centrifugal force became greater than the attraction of the globe; that in this way successive rings of irregular thickness have been thrown off, which each contracted into a planet itself revolving on an axis; and the satellites were formed in the same way from the planets. Geometry shows that this theory remarkably accounts for the facts in the solar system as it was known in the time of Laplace.† Dr. M'Caul considers that it accounts for the earth being spoken of in Genesis as in existence before the sun. He allows that the whole nebulous mass must have been created at one and the same time 'in the beginning;' but he appears to consider

* Sir John Herschel, in his *Astronomy*, and Mr. Grant, in his *History of Physical Astronomy*, both standard works, make no mention of it. Humboldt, in his *Cosmos*, in the passages referred to by Dr. M'Caul (i. 85, 90, and iv. 163), cannot properly be said to 'take it for granted.' His language in the first and second passages is: '*If* the planets have been' so formed; and, in the third passage, he speaks of the primitive internal heat of our planet, 'generated *possibly* by the condensation of a rotating nebulous ring.' The latest difficulty which has appeared, viz. that the satellites of Uranus revolve the wrong way, appears fatal to the theory.

† Professor Huxley says, that Kant first propounded this hypothesis in 1755. See *Lay Sermons, Addresses, and Reviews*, p. 263. Laplace was born in 1749.

that the vapoury ring, out of which the earth was formed, according to the hypothesis, became a globe sooner than the central mass did. There is no reason whatever for this. Indeed, the central mass would have been a globe, a globe of continually diminishing dimensions, from the beginning, and through all the changes; whereas the nebulous ring could not have become a globe for long after the central mass had contracted within the limits of the earth's orbit. The heat which the central mass would radiate away (the loss of which is supposed to have produced the contraction of its dimensions) would keep up the heat of the nebulous ring, and thus retard its contraction into its present dimensions. Again, the theory suggests nothing whatever regarding the luminosity of the nebulous matter. The hypothesis is merely a mechanical contrivance for suggesting how the peculiar motions and positions of the bodies of the solar system may possibly have been acquired from a primitive cause. The Sun's being furnished therefore with a luminous atmosphere on the fourth day, as supposed by Dr. M'Caul, is not taken from this theory, but is borrowed from Genesis, which it is his object to explain. Again, he explains God's making the greater light to rule the day, to mean His giving to the opaque mass of the sun a luminous atmosphere. But the moon is spoken of in Genesis in precisely the same language as the sun, except that it is only of less intensity. But the moon has no luminous atmosphere, but shines only by borrowed light. Again, he uses the same nebular hypothesis to explain the fact which geologists have announced, that none of the fossil plants or animals indicate the existence of climatic distinctions during any age previous to the present or human period. He assumes that the whole of the palæozoic, secondary, and tertiary fossils were created, flourished, and perished during the first and second days of the creative week, that is before the sun had become the centre of light and heat to the earth, and therefore before the present seasons and varieties of climate had begun ; that the works of the third and following days were a preparation of the earth for the reception of man ; and that the plants and animals then created are the progenitors of the present flora and fauna.* But, in reply to this, I would say, that till the third day there was no visible dry land, and yet in the tertiary deposits, at least, there are multitudes of fossils of terrestrial animals and trees. Then, lastly, what were 'evening and morning' on the fourth and following days on this hypothesis, if they were not the ordinary diurnal evening and morning? And if they were so, is there not a want of homogeneity in the interpretation of this formula, which occurs six times in the narrative, as Dr. M'Caul expressly asserts that 'the first three days were not measured by the interval between sun-set and sun-set, *for*, as yet, the sun [according to his hypothesis] was not perfect, and had no light?'

Professor Challis adopts the idea that the days

* *Aids to Faith*, pp. 217-219. At p. 217, l. 6, he says that the first two days 'may' include the primary, secondary, and tertiary formations. But, in the last line, he assumes the truth of this when he says that, the dry land appearing on the third day, the tertiary world was buried by the rise of the ocean.

were periods, and supposes that the sun was created in the beginning, but was prevented from illuminating the earth during the first three periods by a vast stratum of vapour, which vapour he imagines to have been luminous on the first, second, and third days, but to have been in such a state of disquietude during the darkness which separated them, when preparation was taking place for the work of the following day, as to have lost its luminosity, and so to have produced the nights between the periods.* On the

* *Creation in Plan and Progress*, pp. 19-40. The theory of this book is, that in the portion of Scripture we are considering, God gives a Plan of Creation, and that in the subsequent portion and in nature we have a representation of Creation in Progress; and that it does not necessarily follow that the progress has followed the plan in all precise particulars. Professor Challis makes much of Genesis ii. 4, 5, as rendered in the English version and in the LXX, which seems to teach that God made plants, &c., in one sense (in plan) before He made them in another (that is, actually in the earth). But these verses may be better translated. The word rendered in our version 'before,' means also 'not yet,' as in 1 Sam. iii. 7. These verses will then read as follows :—' These are the generations of the heavens and of the earth when they were created. In the day that the Lord God made the earth and the heavens, then no plant of the field was yet in the earth, and no herb of the field had yet grown; for the Lord God had not caused it to rain upon the earth, and there was not a man to till the ground.' This entirely cuts away the ground from under this interpretation.

If, indeed, we retain the present English translation of the passage, which has the support of the Septuagint, it by no means implies that God designed one thing and in the act of Creation produced something different to it. But it rather implies that God created the vegetable kingdom in the stage of plants, herbs, trees— all containing in themselves the power of reproduction when the proper influences of moisture and heat came to act;—and that He did not create it as seed only in the first instance. And similarly He created man and woman in their *adult* condition, and not in some earlier and more helpless stage of existence, through which their offspring would pass.

fourth the vaporous clouds break up and the sun shines forth in its brightness, and henceforth becomes the regulator of night and day as at present. Hitherto the 'evening and morning' have been the alternations of darkness and light, as the thick nimbus thrice lost and twice regained its brightness. But, although the sun rules throughout the remainder of the creative week, the 'evening and the morning,' for consistency sake, is not considered that 'evening and morning' which the setting and rising of the sun once in twenty-four hours actually produced, but something else,* which, however, the theory does not explain, for no notice is taken of the formula 'evening was, and morning was' on the fourth, fifth and sixth days. Here is again the same want of homogeneity in the explanation of this formula as before, and which of itself seems to me fatal to these theories of interpretation.

Dr. Dawson, another advocate of the period scheme, has a theory regarding the creation of man and the dominion given to him over the creatures, which he thinks greatly strengthens the hypothesis of periods. He asserts that carnivorous animals are not mentioned, but *behêmahs*, or herbivorous animals, among the creatures over whom Adam was given dominion. He supposes that Adam was created at the end of the sixth period, along with a group of creatures adapted to contribute to his happiness and having no tendency to injure or annoy, and that they are alluded to in Gen. ii. 19, 20, and that Eden was the region which they inhabited. But that in the regions around, and all over the earth, were the descendants of animals

* *Creation in Plan and Progress*, p. 41.

created in the earlier times of that period, and not all endowed with the harmless dispositions of the latest groups. The cursing of the ground for man's sake, on his fall from innocence, consisted in the permission given to the predaceous animals and the thorns and the briers, of other centres of creation, to invade man's Eden; or, in his own expulsion, to contend with the animals and plants which were intended to have given way and become extinct before him. He assumes that many animals, contemporaneous with man, extend far back into the tertiary period: and that these creatures, not belonging to the Edenic centre of creation, but introduced in an earlier part of the sixth day, are now permitted to exist along with man in his fallen state. He considers that this view strengthens the probability that the creative days were long periods, and opposes an almost insurmountable obstacle to every other hypothesis of reconciliation with geological science.* But Dr. Dawson's assertion, upon which his theory stands, that the word *behêmah* (used in Gen. i. 26) means herbivorous creatures exclusively, is not true. Indeed, in v. 30, where grass is assigned to the beasts of the earth, the word *behêmah* is not used, but the other word, which Dr. Dawson takes to mean carnivorous creatures. Moreover, *behêmah* is the word used in Deut. xxviii. 26, which leaves no doubt that it does not mean herbivorous creatures exclusively: in Prov. xxx. 30, it is used in reference to a lion.

Dr. McCausland is another follower of the period

* *Archaia, or Studies of the Cosmogony and Natural History of the Hebrew Scriptures*, pp. 215–222.

theory. In his book, *Sermons in Stones*, there is much in common with others who take that view, and which therefore need not be alluded to here. I will notice some things which are peculiar to him, and which show the difficulties under which his theory labours, in addition to those which are common to the whole class. He finds that in the Silurian and Devonian strata there are three classes of submarine animals: (1) zoophytes and bivalve mollusks, without visual organs, and which may have existed before light; (2) the higher classes of mollusks and crustaceans, furnished with organs of sight, and which must have come into existence after the creation of light; and (3) the still higher class of vertebrate fish. The first class, he asserts (though nothing of the sort is stated in Genesis), were created on the first day, before the dawn of light; the next on the second day, after the appearance of light; and the third, contemporaneously with land vegetation, on the third day. How can he venture to make these assertions, when it was on the fifth day or period, and not sooner, that 'the waters brought forth the moving creature that hath life?'* He finds it all in the second verse, 'And the Spirit of God brooded upon the face of the waters,' and brought into existence these marine

---

* Dr. McCausland makes much of the Hebrew word *sheretz*, translated 'moving creature' (Gen. i. 20), and says it should be *reptile*, as the verb means to creep. He thus excludes fish from the fifth day, and admits only reptiles, and hence saurians. But the word also means to *swarm*, and, being applied to Pharaoh's frogs (Exod. viii. 3), which did not creep, but hop, must have a more general meaning. Gesenius's meanings are—1. To multiply or propagate itself abundantly. 2. To creep, crawl, swarm.

creatures. He explains the fact of nothing being said of these creations in Genesis, by adding that as this work was spread over, as he assigns it, three days, and could not be appropriated to one only, it was not more particularly mentioned, and, in fact, 'ought not to have formed part of the Mosaic narrative!'* Dr. McCausland follows Mr. Miller in regarding the whole as being a vision, and observes that Moses could not *see* or *hear* what was going on beneath the waters, and therefore this intimation, of the brooding of the Spirit on the surface, was given instead.† How, then, did Moses see the creatures which 'the waters brought forth abundantly after their kind' (v. 21)? Dr. McCausland draws out more forcibly than any of his co-theorists the correspondence between the work of the third day, (when the earth, enveloped in a vapoury atmosphere, and not fully shone upon by the sun, was clothed in vegetation,) and the existence of the coal-fields; and the unwary reader may, doubtless, be carried away into admiration of the parallel so eloquently described. But what are the contents of these fossil coal-fields, even on Dr. McCausland's own showing? Half, at least, are of the fern or bracken tribe, and the rest cryptogamic and flowerless plants.‡ Where, then, is the 'fruit-tree yielding fruit after his kind,' described in Gen. i. 11? How can so minute a depictor of striking coincidences have passed by this fatal discrepancy? 'The evening and the morning,' too, meets with only a passing notice, as it evidently must stand in the way in the period theory under every

---

* *Sermons in Stones*, 1857, pp. 154, 155, 170.
† Ibid. p. 156. ‡ Ibid. p. 166.

form. In one place, indeed, he seems almost to acknowledge, that the effect must have been produced by the sun's alternate rising and setting, when he speaks of the 'dim diurnal light and nocturnal darkness alternately pervading all.'* If this be the case, it is difficult to reconcile it with his theory of periods.

Mr. Warington, in his recent book, *The Week of Creation*, must be regarded as an upholder of the period theory. He regards this portion of Scripture as history, and not a vision or mere ideal representation; he repeatedly speaks of it as 'a history of creation:' he writes of it as 'a professed revelation of otherwise unknown natural facts, whose narration as facts is an essential part of the purpose in view' (p. 5). He says, that it contains 'information of a scientific character, not otherwise obtainable by those to whom it was given;' but 'to impart such information' was not 'its proper end' (p. 23); 'the aim of the narrative was not to enlarge men's views of nature as such, but, through nature, to teach them concerning nature's God' (p. 27). 'But we have a right to expect that the truths taught shall be in harmony with the results of science .... and that the facts alleged as embodiments of these truths shall be really facts, described in language phenomenally correct' (p. 28). He considers, that 'in the six-days' work is to be included the entire history of creation' (p. 65); and that 'these days are without doubt literal days and not long or indefinite periods' (p. 67); but yet that we are not to conclude that it was the

* *Sermons in Stones*, p. 186.

intention to teach us 'that in six literal days of twenty-four hours each the whole of creation was accomplished from beginning to end' (p. 67). In reference to this paradox he adds, 'It by no means follows, that because the *description* speaks of literal days, therefore the *realities* they described were also days' (p. 67). This contradiction he attempts to remove in the following way. We read in Scripture of God's arm, God's eye, God's mouth, the words *used* mean literally arm, eye, mouth, and nothing else. But the *use* here made of these words is figurative. This is a distinction which we can clearly understand. So he says that we read that 'God went down to see,' that 'God smelled a sweet savour,' and that 'God repented:' and he adds that the words used mean literally, went down, smelled, and repented. 'Yet we do not ascribe,' he adds, 'any one of these actions literally to God, but we assert that there were actions of God having the like relation to His nature, which these actions, taken literally, have to our nature' (p. 69). All this is intelligible. Causes, or instruments, or actions are put for their effects. Man's arm, eye, mouth represent his strength, knowledge, directing power. So, though in a degree and sense superhuman, the Divine strength, knowledge, and commands are represented by the Divine arm, eye, mouth, as if the Deity were really possessed of those parts. This idea he applies thus to the case in hand. The six days are literal days; but as the terms occur in a description of the Divine acts, they may really represent something else. This I will admit is possible. But *what* else? Analogy would say, something similar to what would be an *effect* of six days in human life and practice. But in Mr. Warington's explanation the thing represented is Other Periods of time, only of unknown length, periods long enough for the terrestrial work which geology shows us has been done in a practical mundane way in ages past. The six days represent not any attribute of the Deity, as the arm, eye, and mouth do, or even Time in the abstract; but periods of time, differing from the days merely in being of unknown duration. The analogy, therefore, altogether breaks down. That days might represent periods, I admit, if no valid arguments from other sources exist opposed to the idea—which, in fact, I have already shown there are. But then the historical character of the whole passage, which Mr. Warington attributes to it, must be abandoned. The description becomes a mere dramatic representation, and not a history of facts.

I might quote the works of other upholders of the period theory; but the references already made will suffice, as they will represent the views of this class of interpreters, and show, I think, the danger of trusting to any such theories of interpretation. It is surprising what analogies an ingenious mind may trace in things which really have nothing in common. By passing by everything which is contrary, and pressing into service everything which is favourable, writers may persuade themselves and the incautious into a belief of almost any theory so constructed. The ready writer, whose work, *Sermons in Stones*, I have been noticing, comes, apparently, under this description. He brings forward several remarkable analogies between the Mosaic account and the records of the earth, and sets them forth in a very striking manner; but he passes by, unnoticed, with one exception* (which is not, however, met), whatever is opposed to his conclusions. His analogies, moreover, do not prove an identity. There is nothing at all improbable in the idea that the Almighty, in calling into existence plants and animals, at the time that man was created, to be the progenitors of species now existing, should follow in some respects the order of previous creations.†

No such difficulties as those which beset the period theory stand in the way of our reception of the first *The natural-day theory accepted.* method of explanation, in which the days are taken to be natural days. Mr. Miller, though he became the advocate of the period theory towards the close of his life, originally maintained the natural-day theory. The argument which turned him from his former view appears to be this: That an examination of the species of the tertiary strata and of the present flora and fauna indicates, that there has been no such break in organic life preceding the present order of things, as is represented according to this interpretation in Gen. i. 2—no such 'chaotic gulf of death and

---

* This is, the trilobites with eyes have been subsequently found lower down than the sightless zoophytes: see pp. 149, 150, and compare this with p. 148 of Dr. McCausland's book.

† Since writing the above I have met the same sentiment in Dr. M'Cosh's work, *The Supernatural in Relation to the Natural:*—'Who will venture to affirm that God, who has proceeded from the beginning in our Cosmos according to the method of type, that is, model or exemplar . . . . may not have proceeded by type likewise in that necessarily wonderful transaction which ushered man upon the scene? . . . . The account in Genesis may thus be a description of six literal days, as representative of six epochs.' Pp. 343, 344.

darkness."* M. D'Orbigny has, however, re-examined the fossils, and asserts in his *Prodrome de Palæontologie*, that not a single species, either vegetable or animal, is common to the tertiary and the human periods: and therefore that, in his opinion, a break did occur previously to the human period, since it is through species, and species alone, that an hereditary succession is kept up.† This conclusion has since been called in question. All, then, that can be said is, that when eminent men differ so materially on the subject, it shows how uncertain their conclusions must be, and how unworthy of being brought into competition with the plain statements of Scripture. The very fact that eminent men differ in opinion, as to the identity of species of these living and fossil creatures, of itself convinces the impartial bystander, that their rules cannot be so free from uncertainty as some would have us take for granted. Certain canons are laid down in natural history, sometimes arbitrary in their character, which have been so long unhesitatingly received, because nothing has arisen to shake confidence in them, that they are regarded as true; and if a doubt arises, suggested from the side of Scripture, its evidence is regarded as utterly untrustworthy. But there are many instances in which scientific men have been compelled to abandon canons which they have once firmly held. This subject of the exact identity of species of living creatures, and of others seen only in the fossil, and therefore in some degree imperfect, state, is one which

---

* *Testimony of the Rocks*, p. 122.
† See Lardner's *Museum of Science and Art*, vol. xii., Pre-Adamite Earth.

seems to be open to serious doubt.* Moreover, as it is clear that there are many new creatures which came into existence with man and never appeared before, why cannot it be believed that some species which existed in former ages were reproduced? If apparent identity seems indisputable, I cannot see why we should limit the Almighty in His choice of what He should see right to produce when He refitted the world, after its chaotic condition, for the reception of man. This whole subject of species has undergone great changes of opinion, and is still in an unsettled state. Even among living creatures a division of species has often been made which has been erroneous. It is now universally admitted that the cases are extremely numerous in which diversities of age have led to the establishment of species which have no existence in nature; the former, thus distinguished, being those of the same species in different stages of growth.† The cases I have here referred to are instances of species considered at first to be different, which have been afterwards thought to be the same.

* For example : Professor Huxley writes, 'No one would hesitate to describe the pouter and the tumbler as distinct species, if they were found fossil, or if their skins and skeletons were imported, as those of exotic wild birds commonly are—and, without doubt, if considered alone, they are good and distinct morphological species. On the other hand, they are not physiological species, for they are descended from a common stock, the rock-pigeon .... the pouter and the tumbler breed together with perfect freedom, and their mongrels, if matched with other mongrels of the same kind, are equally fertile.' — *Lay Sermons, Addresses, and Reviews,* p. 298, 299.

† *Cyclopædia of Anatomy and Physiology,* art. 'Varieties of Mankind,' p. 302. The following is a recent example of what I have said above : 'In a paper before the Zoological Society, Dr. Günther, in dealing with the clupsoids of the British coasts, gave it as his opinion

I do not know of any illustrations of the opposite kind, viz. where species once thought to be the same are found now to be different. But my argument is, that the question of species is an unsettled one, and, therefore, the identification of fossil and living species open to uncertainty.

In adopting the explanation that the days were literally days of twenty-four hours, we have but to suppose that an interval of untold duration occurred between the first creation of the heaven and the earth, (that is, of the planetary and starry heavens and the heaven of heavens and all that they contain, and of the earth, as a member of the solar system), and the preparation of the earth for the reception of man; that in this interval the plants and animals, which we find fossilized in myriads in the earth's crust, lived, died, and were entombed, to tell, in after ages, their own story; and that, regarding these —with which man was not concerned—the Scriptures are silent.* Thus, the three geological discoveries regarding the Antiquity of the Earth, the Existence

that the whitebait is really a young herring. We are glad to learn the belief of one of the most eminent of European ichthyologists, and the more so as it confirms the opinion expressed in an article in one of our earlier volumes, in which the writer expressed his conviction that the anatomical affinities of the herring and whitebait were so close as to justify their being grouped into one species.'—*The Popular Science Review*, No. 29, October 1868, p. 456.

* 'There is nothing to connect the time spoken of in Gen. i. 2, with that of the first great declaration of the creation of all things *in the beginning*. "In the beginning God created the heaven and the earth." Rather, of the forms of speech which could have been chosen to express past time, *that* has been chosen which least connects the state, when the earth was one vast waste, with the time when God created it. Both were in past time; but there is nothing to connect

of Animals and Plants long prior to the appearance of Man, and the Existence of the Sun also prior to the work of the six days, may be true, and yet find no opposition in the statements in the book of Genesis, interpreted according to this theory, which takes the days to be natural days; and Scripture and Science are found to be not at variance. The six-days' creation exhibits a series of creative acts, which terminated in the appearance of the Human Race upon the scene. The animals and plants then created were the progenitors of those which now, possibly with others since created, tenant the earth.

Mr. Goodwin has attempted to cast upon this interpretation the reproach that it teaches nothing. What! have these sublime verses taught nothing from age to age since they were revealed? Are the lessons they convey dependent upon confirmations, supposed or real, which we, in these later times, may find in the opened book of nature? Is this the spirit in which we are to receive a message from on high,

those times together. First, we have, as far back as thought can reach, creation, *in the beginning*, of all those heavens of heavens through those all-but-boundless realms of space, and of our earth. Then, detached from this, a past condition of the earth,—but not a condition in which God, who made all things very good, ever made anything. What follows is connected with this state. First, we have a contemporary condition (as it is expressed in Hebrew), "and darkness upon the face of the deep;" then a contemporaneous action, of more or less duration, "and the Spirit of God brooding upon the face of the waters;" then successive action (as this, too, is expressed in Hebrew), "And God said;" which is continued on through the rest of the history of the Creation. It seems, then, that God has told us, in the two first sentences, just what concerned us to know: first, that He created all which is; then, how He brought into order this our habitation which He has given us. What intervened between that in which the Almighty deigns to reveal to us that He is the Great Creator of all things in heaven and earth? I have, however, already pointed out some certain truths which this communication does teach. This opening portion of the book of revelation appears to have been written to communicate a right view of the origin of the universe, as an antidote to those false notions which had already arisen in the time of Moses, or would afterwards arise on that subject. The leading principle, which the first verse teaches, is, that the universe exists, not independently of God, by any necessity or by any inherent power; nor yet contemporaneously with God, as being co-existent with Him; nor yet in opposition to God, as a hostile element; but dependently upon Him, subsequently to Him, and in subjection to Him. This leading principle of the subordination of matter to God in every respect is expressed in broad terms in these opening words, 'In the beginning God created the heaven and the earth.' The same great principle

creation "in the beginning" and that re-modelling for our habitation, does not concern us; and on this God is silent. He tells us the first and the last, that He created all things, and that He prepared this our beautiful earth for us, and created all things in it and ourselves. In the interval there is room for all the workings of God which geology speaks of, if it speaks truly. This history of Creation in Genesis falls in naturally with it, in that it *does* say that this our mysterious habitation, which God has made the scene of such wondrous love, was created "in the beginning," *i. e.* before the time of which it proceeds to speak. Another period of undefined duration is implied by the words, "And the Spirit of God was brood*ing* upon the face of the deep." For action, of course, implies time in which the action takes place. And this action was previous to that of the first "day" of the Creation, which begins, like the rest, with the words, "And God said."'—Dr. Pusey on *Daniel*, Preface, p. xix.

runs through the beautiful description which follows of the orderly preparation of the earth to become the residence of man, God's rational creature whom He was now about to call into existence. Matter owes its form and modifications to the will of God. In itself dull and inert, it receives its vivifying capacities from the influence of the Spirit of God brooding over the deep : (i. 2). The progressive improvements in its condition were the direct effects of God's will. No interposition of secondary causes is recognized. 'He spake, and it was done' (Ps. xxxiii. 9). And the pointed terseness and sharpness with which the writer sums up the whole transaction in the three expressions, 'God said'—'It was so'—'God saw that it was good,' the first declaring the Divine volition, the second the immediate result, the third the perfection of the work, harmonizes well with the view he intended to express. Thus the earth became in the eyes of the pious Hebrew, and is seen by us also to be, the scene on which the Divine perfections were displayed: the heavens (Ps. xix. 1), the earth (Ps. xxiv. 1 ; civ. 24), the sea (Job, xxvi. 10 ; Ps. lxxxix 9 ; Jer. v. 22), 'mountains and all hills, fruitful trees and all cedars, beasts and all cattle, creeping things and flying fowl' (Ps. cxlviii. 9, 10) all displaying His goodness and power.* The lesson here taught us is not

---

* Compare Dr. Smith's *Dictionary of the Bible*, article 'Earth.' The following observations are from Bishop Wordsworth's comment on Gen. i. 1 :—'*God created.* Here is a prophetic protest against the false systems of Natural Philosophy which have prevailed in the world. *God created all things.* The world is not God, as the Pantheists affirm. It did not exist from eternity, as the Peripatetics taught. It was not made by fate and necessity, as the Stoics said. It

a scientific lesson, but a moral one: and therefore to attempt to deduce from it a scientific history of the Earth is altogether out of the question. The data for this purpose are not given. If we are tempted to regret that we can gain no precise scientific information from Genesis regarding the details of the original creation, we should resist such a temptation, and call to mind the great object of the Scriptures— to tell man of his origin and fall, and to draw his mind to his Creator and Redeemer. When the Almighty speaks of the works of His hands, it is with a majesty and dignity which become the Maker of the world. He speaks in language which declares Him to be the originator of all things, the Almighty Lord of heaven and earth. He condescends not to describe the process or the laws by which He worked: all this He leaves to reason to decipher from the phenomena His world displays.* My design in attempting, under the first method of interpretation,

did not arise from a fortuitous concourse of atoms, as the Epicureans asserted; nor from the antagonism of two rival powers, as the Persians and Manichæans affirmed; nor was it made by Angels; nor by the emanations of Æons, as some of the ancient Gnostics held; nor out of matter, co-eternal with God, as Hermogenes said; nor by the spontaneous agency and evolution of self-developing Powers, as some have affirmed in later days. But it was created by One, Almighty, Eternal, Wise, and Good Being—GOD.'

* '. . . . The first chapter of Genesis stands alone among the traditions of mankind in the wonderful simplicity and grandeur of its words. Specially remarkable—miraculous it really seems to be— is that character of reserve which leaves open to reason all that reason may be able to attain. The meaning of those words seems always to be a-head of science,—not because it anticipates the results of science, but because it is independent of them, and runs, as it were, round the outer margin of all possible discovery.'—Duke of Argyll, *Primeval Man,* p. 367.

to give a physical representation of what the process of the six days' work might have been, is not to impose this or any suchlike meaning upon the words, but to show that the language is in no way contradictory to scientific possibilities. It rather detracts from the simple grandeur of the whole, and diverts attention from the great lesson to be conveyed, to seek for a scientific meaning, especially, too, where it was not designed. I take this meaning, not as necessarily taught by Scripture, but as merely illustrating it in those scientific points. I receive it full ready to change it for another, if scientific study require it, and the language of Scripture have no unnatural interpretation forced upon it. We may well rise, indeed, from the contemplation of the sacred volume with admiration at the wisdom with which its phraseology has been chosen; so that while human systems have disappeared before the light of advancing knowledge, Holy Scripture, in its original tongues, remains pre-eminent, and no charge of error brought against it has ever been substantiated.

I cannot close my remarks on this part of my subject without entering a protest against the practice of propounding, in support of Holy Scripture, imperfect theories, derived from a very partial knowledge of the past history, or even the present laws, of the physical world. Mixed as these theories generally are with conjectures and ill-determined facts, they become most dangerous to the stability, not of Scripture, which is an immovable rock, but of the faith of those who are carried away by them. In what I have written I have appeared

Danger of adopting imperfect theories.

perhaps to lean upon the recent researches of M. D'Orbigny. My object in doing so is to meet the upholders of the period theory on their own ground. But I am aware that further research may lead to a modification of his results, and that the doubts regarding them may, in some measure, be confirmed. Surely, much as we may justly admire the pursuit of Science, we have learnt caution ere this not to receive its results too hastily, as the established expression of the laws of nature. No principle seemed more sure than that which Sir Charles Lyell announced some years ago in an early edition of his *Principles of Geology*, that in many respects the species of the tertiary beds are the same as in the present flora and fauna. He gave the name Eocene, Meiocene, Pleiocene, and Pleistocene * to express this idea, and to present the gradual increase of recent species in the tertiary beds as they approach the human period. But this has been controverted by M. D'Orbigny, in his more recent and extended researches, which seem to show that there is not one species in common in the tertiary and present plants and animals. Sir Charles Lyell has, in his last work, alluded to M. D'Orbigny's conclusions; but re-asserts his former views, based originally upon the investigations of M. Deshayes.† I can but repeat, that, if eminent men are so divided in their opinion on these matters, it

---

* These words mean, Dawn of recent, Minority of recent, Majority of recent, Greatest number of recent (species).

† 'In the year 1830, I announced, on the authority of M. Deshayes, that about one-fifth of the mollusca of the Falernian, or Upper Meiocene strata of Europe, belonged to living species. Although the soundness of that conclusion was afterwards called in question by two or three eminent conchologists (and by the late M. Alcide D'Orbigny among

shows how uncertain their conclusions must be, and how unworthy of being brought forward as arguments against Scripture. The science of Geology—one of the most ennobling studies of the day—is in far too young and unsettled a state to justify us in bringing its results into competition with the brief and unscientific, though literally true, description of God's work of creation in the Book of Genesis.

I cannot do better than refer to an admirable article in the *Universal Review* (July, 1859) on 'Illogical Geology,' in which the unavoidable inconveniences of a tendency to rash generalization are well pointed out, while the necessity of exercising the faculty, if Science is to advance, is made equally clear. We must speculate in order to make progress; but it is only well-matured results, demonstrated beyond dispute, which must be allowed to take the position of established theories. The writer draws a striking lesson from the history of astronomical science. There followed one another, beginning from the earliest times, five provisional theories of the Solar System, each in its turn held as final, till the sixth and absolutely true one was reached. In these five theories may be traced both the tendency men have to leap from scanty data to wide generalizations, which are either untrue or only partially true, and also the

others), it has since been confirmed by the majority of living naturalists, and is well borne out by the copious evidence on the subject laid before the public in the magnificent work edited by M. Hörnes, and published under the auspices of the Austrian Government, *On the Fossil Shells of the Vienna Basin.*'—See Lyell's *Antiquity of Man*, p. 430. On this debated matter of species we may soon have as strong opinions advanced on the other side.

necessity there seems to be for these premature generalizations as steps to the final one. The same laws of thought have prevailed and are prevailing in the younger science of Geology. We have had crude utterly untrue dogmas, for a time, passing current as universal truths. We have evidence collected in proof of these dogmas. By-and-bye a colligation of facts is produced in antagonism with them. Eventually, a consequent modification is suggested. In conformity with this somewhat improved theory, we have a still better classification of facts; a greater power of arranging and interpreting the new facts, now rapidly gathered together; and further resulting corrections of theory. We are at present in the midst of this process, and therefore it is not possible to give an adequate account of the development of geological science; the earlier stages are alone known to us. As one source of uncertainty in some of the general results at which geologists are supposed to have arrived, we may mention the difficulty of determining the relative age of strata which are not contiguous. Neither their mineral structure nor the fossils they contain by any means lead always to conclusions which are beyond dispute. This is a matter of great importance; for much depends upon it.* Another is the metamorphic effect of heat. Sir Charles Lyell says, that in some cases every vestige of vegetable and animal remains in limestones and in clay formations has been entirely obliterated by this process. It may, then, so happen that the fossils now in existence are but the last chapter of the earth's

* Let the reader consult the Anniversary Address before the Geological Society in 1862, by Professor Huxley.

history, and that many primary chapters, stretching back to a time immeasurably more remote even than existing fossils, have been *burnt*, and with them all the records of life we may presume that they contained! The analogy which our period theorists draw between existing fossils and the account in Genesis would be, in this case, altogether thrown out. The subject is too vast and too unsettled, to allow us to base any trustworthy conclusions upon such comparisons. I have no desire to disparage the study of this noble science; but I would rather promote it. But when its results, which are, after all, only approximations, and often very uncertain approximations, to the true history and condition of this wonderful globe on which we live, are turned into arguments against the inspiration of the Holy Scriptures, I stand amazed at the temerity of the men, who, knowing anything of their relative claims, will doubt, for one instant, on which side the error must lie, if any real discrepancy is found to exist.

4. Besides the three points of apparent difference between Scripture and Science which I have been considering at such length, geology gave rise to another formidable difficulty regarding the existence of Death in the world before the fall of Adam. The myriads of creatures which the strata have brought to light lived and died ere Adam came: and yet St. Paul has said, 'By one man sin entered into the world, and death by sin' (Rom. v. 12). So unanswerable has this objection appeared to some, that blindfold they condemn the whole science of Geology, and ignore the universal testimony of the greatest and

[margin: Death before the Fall.]

best men. And no doubt, while ignorant of the facts which the Book of Nature reveals, we should conclude from the Apostle's words that it was the sin of Adam that had brought Death upon the irrational as well as the rational creation. But is this the necessary meaning of the passage? By no means. Science here comes to our aid to correct the impressions we gather from Scripture; and the lesson we learn from the Apostle is, not that Death had never appeared, even in the irrational world before the Fall of Man, but that in that fearful event sin had degraded God's intellectual creature to the level of the brutes in his animal nature, and in his spiritual to that of a lost and fallen being. Death received its horrors when its sentence fell upon man, who alone was made in the image of God.*

* Two hundred years ago—long before the science of Geology called for the belief that mortality had been stamped on creation and had manifested its proofs in the animal races previously to Adam's appearance—Jeremy Taylor could write as follows regarding Adam himself before the Fall. He considers him to have been created mortal—not merely liable to become mortal, but actually mortal.

'For "flesh and blood," that is, whatsoever is born of Adam, "cannot inherit the kingdom of God." And they are injurious to Christ, who think, that from Adam we might have inherited immortality. Christ was the giver and preacher of it: "He brought life and immortality to light through the Gospel."' Again:—' For that Adam was made mortal in his nature, is infinitely certain, and proved by his very eating and drinking, his sleep and recreation,' &c.—*Works of Jeremy Taylor*, by *Bishop Heber*, vol. ix. pp. 74, 76.

And in another passage quoted by Professor Hitchcock:—' That death which God threatened to Adam, and which passed upon his posterity, is not the going out of this world, but the manner of going. If he had stayed in innocence he should have gone placidly and fairly, without vexatious and afflictive circumstances; he should not have died by sickness, defect, misfortune, or unwillingness.' These senti-

5. Another difficulty, which the progress of scientific discovery has originated, arises from the theory of Specific Centres. According to this discovery, every species, whether of plants or animals, is confined to a certain region or habitat, beyond the boundaries of which its individuals cannot live. Each species, therefore, must have diverged from some centre within its region; and this is called a Specific Centre: and these centres must have been the foci of creation. No doubt the boundaries of these regions may have varied since the six days' creation, under the influence of climate. But it is contended that no change of climate which is likely to have occurred, can account for the transfer of the centres to such considerable distances as many

[margin: Specific centres.]

ments I quote, not as necessarily approving them, but to show that so good and learned a man as Jeremy Taylor had a view regarding death and mortality no less unusual than that which geology demands.

'It is certainly a startling fact,' writes Bishop Ellicott, 'that ages before the sin of man cast the shadow of vanity on the world, suffering, in one of its forms, the corporeal, was certainly present. The very stones and rocks bear witness of it, the acknowledged presence in the pre-Adamite world of the fierce and fell race of the carnivorous animals renders its past existence a certainty. . . . In every endeavour to view suffering in its most comprehensive and general aspects, we must be especially careful to draw a clear line of demarcation between the corporeal sufferings of the individuals that belong to lower genera unendued with foresight and reason, and the mixed *mental* and corporeal sufferings of a personal and intelligent being, the immediate child and offspring of God. . . . The scattered hints and speculations of earlier writers, afterwards more fully developed by some of the deeper thinkers of the seventeenth century, that regard the early history of the world and the fall of angels as in *some sort* of connexion, are certainly not wholly unworthy of our consideration. . . . How far the disturbance caused by that fearful lapse was propagated through the other realms of creation, we know not. How far demoniacal malignity might have been permitted to introduce or multiply sufferings into the

of them are found to be from the limits of Paradise. This appears to be at variance with the account in Genesis, which seems to represent the creation, of animals at least, to have been in regions within the reach of Adam. But this difficulty need not stagger us, unexpected as it is. For in the first place, it is not impossible that the regions, of which the limits are far from the boundaries of man's first residence, have become the scenes of creative power at epochs subsequent to the six days' work. The statement that 'the heavens and the earth were finished, and all the host of them,' does not imply that the Almighty would never exert His creative power again, but that the work of the six days was then completed.* It has been said, that at the antipodes there are animals

early animal world Scripture does not, even incidentally, reveal. Still, it does not seem utterly presumptuous to imagine that there might have been the same powers of evil partially and permissively at work in a pre-Adamite world, that, at a later period, when man's sin had wrought a still more frightful confusion, were permitted to drive the swine down the steeps of Gennesaret.'—See Bishop Ellicott's *Destiny of the Creature*, Serm. ii.

For thoughts upon this view of the subject, I would refer my readers to a suggestive and original, though speculative, article in the *Christian Remembrancer*, vol. xli., first quarter of 1861, p. 402, which is worth pondering over.

* 'The difficulty as to the animals found, each in their several habitats, in Australia, New Zealand, &c., is properly no *scientific* difficulty. It lies on the surface; but it presupposes that the "rest" of God, spoken of in Genesis, implies that He created nothing afterwards, which is contrary to our Lord's words, 'My Father worketh hitherto, and I work,' and to the fact that He is daily and hourly creating those myriads of human souls which He infuses into the bodies prepared by His Providence.'—Dr. Pusey's *Lectures on Daniel*, p. xxii.

Dr. Colenso, in his criticism of a former edition of this work (Part. IV. p. 131), assigns no reason why the words, 'The heavens

apparently approaching extinction, and others really extinct, and that this appears to show that they were created earlier and not later. Were this generally true of the species of those parts, it might be of some importance, but otherwise it shows only that those few species were of short continuance. And further, there is nothing in the account of the six days' creation to militate against the idea that creation may have been going on over the whole surface of the earth at the same time. It simply requires us to suppose that the animals brought to Adam for him to name them, must have been those only in the neighbourhood of Paradise. Indeed, this seems to be the most natural interpretation of the narrative in Genesis, as I will show.

The first three chapters consist of two parts. The first portion describes the week of creation; and in the second, beginning at ii. 4,* the sacred writer pro-

and the earth were finished, and all the host of them,' should imply that the Almighty could never exercise His creative power again. His preconceived notions evidently trammel his mind in his supposing that the first chapter of Genesis necessarily describes everything that the Almighty ever created. With regard to his second interjected quotation, with italics, I would say, that Dr. Colenso appears often to throw emphasis upon wrong words, so as to pervert the direct meaning of the writer, and to draw attention to the wrong point. The correct idea, as it appears to me, conveyed in the words he quotes in brackets, viz. 'The man called names to all the cattle and to the fowl of the heaven, and to every animal of the field,' is, that Adam gave names to creatures, and that he did so to all kinds of creatures; not necessarily that there was not a creature or even a species in the whole world which was not brought to Adam to receive its name.

* It has been a matter of discussion whether v. 4 belongs to v. 3 or is the beginning of a new paragraph. It appears to me that the words, 'These are the generations,' are the beginning of a fresh paragraph which terminates with chap. iii., written by the writer who wrote the account of the creative-week, and that the word 'these'

ceeds to the narrative of man's moral probation and his fall, the key to his subsequent history. In preparation for this additional particulars are given to enable us to understand the place and circumstances of the temptation, and the means which the tempter used to effect his purpose. Man's frail origin, out of the dust, the earthly part of his nature, is therefore here first mentioned; the garden is described in which he was placed when created, with its tree of life, and the tree of knowledge of good and evil; the origin of woman in man's want of a companion, as none of the creatures which passed in review before him met that want, and man's relation to her, are now told us, that we might the better understand their double fall. This all occurs in Eden. It was, therefore, only the animals in Eden that passed in review before Adam to receive names. Animals beyond the limits of Eden are not

refers back to that description. 'These [which I have now described] are the generations of the heavens and of the earth.' To imagine that the writer changes simply because the title of the Deity is changed from 'God' to 'Lord God,' is an arbitrary hypothesis. Moreover, the hypothesis very soon breaks down altogether; for in the very midst of this narrative the name 'God' is used three times in the third chapter. Dr. Colenso has a theory to account for this—a theory to explain a theory, viz. that the sacred writer would not put the name of 'Lord God' into the devil's mouth. But if he, a mere critic, can divine a reason why the name should be changed in this instance, is it not rather self-confident to assume that there is no reason, because he cannot guess it, why the sacred writer should first have used 'God' in his description of the creation, and then 'Lord God' in his description of the temptation and fall? And so in other places in this book of Genesis, wherever a change takes place in the name of the Deity, this critic fearlessly asserts, as others do whom he follows, that interpolations from other writers have taken place. Save us from such plastic ingenuity, which, on the flimsy pretext of a critic's theory, will tear to bits the sacred text which has been handed down to us from age to age.

alluded to in the narrative. Indeed, it is observable that beasts of the field only and birds are mentioned; reptiles, and fish, and the 'great whales,' are not alluded to.*

The difficulty, therefore, which the theory of Specific Centres was supposed to introduce, altogether vanishes.†

There is a remarkable illustration of the truth of these narratives in the first and second chapters of Genesis which is worth noticing here. It is known

* The same view Professors Keil and Delitzsch take in their commentary on the Pentateuch. 'To call out this want [of a help-mate], God brought the larger quadrupeds and birds to man, *to see what he would call them* (lit. each one); *and whatsoever the man might call every living being should be its name* (Gen. ii. 10). The time when this took place must have been the sixth day, on which, according to chap. i. 27, the man and woman were created; and there is no difficulty in this, since it would not have required much time to bring the animals to Adam to see what he would call them, as the animals of Paradise are all we have to think of; and the deep sleep into which God caused the man to fall till he had formed the woman from his rib, need not have continued long. In chap. i. 27, the creation of the woman is linked with that of the man; but here the order of sequence is given, because the creation of woman formed a chronological incident in the history of the human race, which commences with the creation of Adam.' See Clarke's *Translation*, p. 87.

† Dr. Colenso, misled by some of his German masters, has asserted that the accounts in Gen. i. and Gen. ii. are contradictory to each other. I cannot do better than quote from the late Dr. M'Caul's admirable reply, condensing from the last work he wrote ere the Church was deprived of his invaluable services.

1. Gen. i. 9, 10. The land emerges from the waters, and was therefore, Dr. Colenso says, saturated with moisture, which is contrary to Gen. ii. 6. *Answer.*—In Gen. i. 9, 10, it is expressly said that the land was *dry*, not moist.

2. Gen. i. 20, 24, 26. Birds and beasts are created before man, Gen. ii. 7, 19, the opposite occurs. *Answer.*—In this second narrative, the historian alludes to events, not necessarily in chronological order as in the first chapter, but as they bear upon the object which he has

that the cerealia, which produce bread-corn, on which the human race so depends for subsistence, exist only as *cultivated* products of the soil. They perish, as far as concerns usefulness to man, without man's care. In correspondence with this necessity for cultivation, we find, in both these accounts, almost incidentally, and certainly with no direct statement of its necessity, that Adam, even before the fall, is admonished to cultivate the vegetable kingdom. 'And God blessed them ; and God said unto them, Be fruitful, and

now in view. From v. 8, 9, it might otherwise seem that the trees did not grow in the garden till after man was placed in it; whereas v. 15 proves that that was not the case. [I may here say, that the Hebrew has only one past tense ; it has no pluperfect. Hence, v. 8 might just as well be translated, 'And the Lord God had planted a garden . . . and there He put the man whom He formed.' I would also observe that the English reader must not be misled by the words, ' So,' ' Thus,' ' But,' ' Therefore,' ' Now,' ' For,' with which some of the verses in these three chapters begin; for they are all *Vau* in the Hebrew, the word translated ' And ' in all the other verses.]

3. Gen. i. 20, the fowls that fly are made out of the waters ; Gen. ii. 19, out of the ground. *Answer.*—This is a blunder which a Hebrew scholar would not make. In Gen. i. 20, ' fowl,' is not in the accusative but in the nominative, and the words should be, ' and let the fowl fly above the earth ;' nothing is said there of what the fowls were made of.

4. Gen. i. 27, man is created in the image of God ; Gen. ii. 7, he is made of the dust of the ground, and merely animated with the breath of life, and only after eating of the forbidden fruit is it said (iii. 22) that he was become like God. *Answer.*—Here is no contradiction. We are not told in chap. i. what man's body was made of. By the breath of life cannot be meant simply animal life ; for in the narrative before the eating, Adam is described as intelligent, free, moral, lord over other creatures. Is not this to be more than a mere living creature ? Is not this to be like God, to have the stamp of His likeness ?

5. Gen. i. 25, man is made the lord of the whole earth ; Gen. ii. 8, 15, he is placed only in the garden of Eden to dress it. *Answer.*— In the first God speaks of man as of the whole human race and its destiny ;

multiply, *and replenish the earth and subdue it*' (Gen. i. 28). 'And the Lord God took the man and put him in the garden of Eden *to dress it and to keep it*' (ii. 15).*

6. Another instance in which Science has been looked upon as inimical to Scripture, is the assertion, now universally made by geologists, that no known traces exist of the Noachian Deluge.

<small>No known traces of the Deluge.</small>

The disappointment which this has occasioned

---

in the second chapter, the particular circumstances of the individual, Adam, are related.

6. Gen. i. 27, man and woman are created together, and, as is implied, in the same kind of way. Gen. ii. the beasts and birds are created between the man and the woman. *Answer.*—Gen. i. 26-28 indicates that man and woman were not created together, *i.e.* simultaneously. 'And God said, Let us make man' [Adam, without the article]. Here the language is indefinite, and refers to the whole human race. But then follows, ' And God created *the* man [Adam, with the article] in His image; in the image of God created He HIM [masculine]; male and female created He them.' This is perfectly consistent with the more extended narrative of the second chapter.

* In connexion with this I quote the following striking remarks : ' Not a stalk of corn remains where man is not. If corn cannot now live without man's aid, it is an unavoidable inference that man was instructed from the first to cultivate corn. So strong has been the conviction of all ages, that the cereals are not spontaneously produced, that the mythologies of India, Egypt, and Greece ascribe their cultivation to direct Divine interference. The Medes, who were the descendants of Medai, a son of Japhet (Gen. x. 2), and among the earliest of recorded nations, certainly anterior to the Chaldæans, connected their notions of piety with the cultivation of the earth as a duty enjoined on them by God. . . . As Mr. Vicien stated at the meeting of the British Association in Birmingham (1865), no trace of the existence of the cereals can be discovered in geological formations that can be imagined more than 6000 years old.'—*The First Man, and His Place in Creation*, by George Moore, Esq. M.D. p. 310. See also Rev. Hugh Macmillan's *Bible Teachings in Nature*, chapter on Corn.

has been felt all the more severely, because the advocates of Revelation had long been in the habit of pointing triumphantly to the rocks in all parts of the earth as containing *shells* even to the highest peaks, and so being infallible witnesses to the fact of the Deluge. Geologists used to support this view. One of their number, eminent both for his eloquent expositions and thorough acquaintance with the science, had even written a work* upon the subject, describing a cave at Kirkdale, in Yorkshire, where bones of numerous animals had been accumulated, it was supposed, by the waters of the Deluge. But it is now acknowledged by all geologists that these conclusions were premature. In a subsequent work this author publicly renounced his former views upon the subject, and recalled his book. Further consideration has shown, that the Deluge cannot have been the occasion of embedding shells and other fossils in solid rocks, and to such a depth as they are found. Nor are the superficial deposits—those, for instance, in the Yorkshire cavern—such as a temporary deluge could have produced.†

The more the progress of scientific discovery has brought to light the varied agencies which are perpetually at work in changing the aspect of the earth's surface, the more is it seen, that it was unreasonable to expect to find traces of the great cataclysm at the present day, so many ages after its occurrence. Any marks it left must have been long since obliterated, or so mixed up

---

* *Reliquiæ Diluvianæ.*

† See this subject fully discussed in *Testimony of the Rocks*, Lect. 8 —on the Noachian Deluge.

with the effects of subsequent gradual changes as to be undecipherable, even if they ever possessed any characteristic features peculiar to themselves. The marvel of this great historic event was the presence of so vast a body of waters—their rapid appearance and as rapid disappearance—the windows of heaven being opened, and the fountains of the deep being broken up. Whether this great catastrophe was brought about by the intervention of second causes or not, it was by the interposition of the Almighty for the punishment of a guilty world. The record of this Scripture conveys to us; and Science, though robbed of its supposed power of illustrating the Scripture truth, nevertheless places no obstacle in the way of its reception.

7. The history of the Deluge furnishes an illustration of my subject in another way. It has been conceived by most readers of Scripture, that it describes the Deluge as having covered the surface of the whole earth. To this view Science of late years has presented various difficulties.

The Deluge not over the whole earth.

Of course, a believer in the Divine Power can have no difficulty in admitting any miracle, however astounding, so long as it does not involve an impossibility, and is clearly demanded by the sacred narrative. But he must not be charged with scepticism, or with favouring it, if he seek so to interpret the record as to avoid an impossibility; or if he endeavour to confine the miracle within limits proportioned to the occasion; or if he search for an explanation, in part at least, in the operation of second causes—by which the Almighty, in many recorded instances, has worked His wonders: for if we exaggerate the demands of Scripture on men's faith, beyond what the text, fairly interpreted, absolutely requires, we make ourselves, so far, responsible for their scepticism. That secondary causes were made use of, though no doubt in a subordinate manner, in the miracle of the Deluge, is apparent from the language—'the fountains of the great deep were broken up, and the windows of heaven were opened, and the rain was upon the earth forty days and forty nights' (Gen. vii. 11, 12). We cannot be wrong, therefore, in seeking the most simple combination of secondary causes which the Almighty may have brought into play to effect His purpose.

Now Mr. Hugh Miller, in his work, *The Testimony of the Rocks*, has shown how all the phenomena of the Deluge might have been produced by the gradual submergence and rising again of the country comprised within a radius of a few hundred miles around the dwelling-place of Noah, so as to include the portion of the globe then inhabited. This phenomenon of the change of level of large portions of the earth's surface, by depression or elevation, is not unknown to geologists ; though the periods in which these vast oscillations occur are of immeasurably longer duration than that of the Deluge. He shows that the depression during the first forty days might, nevertheless, have been so gradual as to have been imperceptible, except from the effects—the pouring in of the mighty waters from the neighbouring seas into the growing hollow, and the disappearance of the mountain-tops. And when, after a hundred and fifty days had elapsed, the depressed hollow began slowly to rise again, the boundless sea around the ark would flow outwards again towards the distant ocean, and Noah would see that 'the fountains of the deep were stopped,' and 'the waters were returning from off the earth continually.' (Gen. viii. 2, 3.)

This process, miraculous though it be in thus calling into sudden action secondary agents, meets the difficulties of the case in a way which no other known hypothesis will. It supplies and disposes of the mighty mass of water required for the catastrophe: it makes the miracle proportioned to the occasion, viz. the destruction of the human species for their wickedness; and, above all, it limits the number of animals which Noah would have to accommodate in the ark within reasonable bounds. Mr. Miller mentions an interesting calculation made by Sir Walter Raleigh, to show that Noah's ark was capable of holding all the then known animals of the world, with provisions for all the time during which the earth was submerged. The calculation of this great voyager is justly considered to have been sober and judicious. But our growing acquaintance with the animal kingdom has converted his trustworthy result from being an argument for a Universal, to that for a Partial Deluge. The eighty-nine known animals of his time would now embrace but a single region. There are between one and two thousand different species of mammals at present known! To this extraordinary increase in our knowledge may be added the six thousand two hundred and sixty-six birds of Lesser, and the six hundred and fifty-seven reptiles of Charles Bonaparte, or, subtracting the sea-snakes and turtles as fitted to live outside the ark, his six hundred and forty-two reptiles.* Now granting that all these thousands of species of mammals, birds, and reptiles could have been brought from all parts of the earth, and actually assembled round Noah, and afterwards replaced in their respective habitats all over the globe, it seems impossible that they could have been all accommodated within the prescribed limits of the ark during the earth's submergence.

The question has been asked, Why were *birds* taken into the ark, if the Deluge were only Partial? But this objection is based upon an error in natural Science, into which even naturalists of the last century, such as Buffon, not unfrequently fell: viz. that of assigning to *species* wide areas in creation, which in reality they do not occupy. A better acquaintance with the habits of many of the non-migratory birds will convince such an objector, that even in a local deluge, of the extent which we suppose the Deluge may have attained, many species would have become extinct but for their preservation in the ark, as the surrounding regions could not have supplied them.†

But is not this notion of a Partial Deluge contrary to the express language of Scripture? The words of Scripture, were there no facts like those I have mentioned to modify our interpretation, would, by most persons, be understood as describing a Universal flood of waters over the whole extent of the globe: there would be no cause for questioning this, and

---

* *Testimony of the Rocks*, p. 323.   † Ibid. p. 292.

therefore no ground of doubt. But when the new facts become known, as they are at present, then the question is started, Does the Scripture language present any insuperable obstacle to this more limited interpretation? That it does not, may be inferred from the fact, that two of our celebrated commentators on Scripture, Bishop Stillingfleet and Matthew Pool, both in the seventeenth century, long before the discoveries of natural Science required it, advocated this view. The strongest expression in the whole account is this, 'All the high hills that were UNDER THE WHOLE HEAVEN were covered' (Gen. vii. 19). But that, if other circumstances require it, this phraseology may refer solely to the region affected and not to the whole globe, is apparent from the use of the same expression by the same inspired writer in another place, in which it is evident, that he cannot have intended the whole globe, but only Palestine and the countries in its immediate neighbourhood : 'This day will I begin to put the dread of thee and the fear of thee upon the nations that are UNDER THE WHOLE HEAVEN, who shall hear report of thee, and shall tremble, and be in anguish because of thee' (Deut. ii. 25). Compare this with xi. 25, where the extent of the dread is limited to 'all the land that ye shall *tread upon!*'

With some minds the argument of the Divine Omnipotence is sufficient to meet all the difficulties of a Universal Deluge : to others they have appeared so formidable as to recommend the theory of a Partial Deluge,—partial not with reference to the human race, but the surface of the earth—which certainly furnishes a ready and complete answer to all the objections.*

The difficulties which I have been considering in this chapter refer to the earlier parts of the book of Genesis, from which source indeed spring others which I discuss in the next chapter. It is needless perhaps to say to those who have read what has gone before, that I am no advocate for forced reconciliations between Scripture and Science. Scripture does not speak in scientific language ; nor should we desire, as it appears to me, to draw scientific conclusions from its statements, where it does touch upon the phenomena of nature. Obscurities and difficulties on this subject may always be looked for, in consequence of the different points of view from which the natural world is regarded in Scripture and by the scientific investigator. What I contend for—and this the very title of my book indicates—is, that, though many charges of variance between Scripture and Science have been made, not one has been substantiated. Where simple explanations of apparent difficulties can be given, it is satisfactory. But were none forthcoming, nothing is necessarily proved against Scripture—as is too generally supposed ; but our ignorance is brought to light. In such cases forced reconciliations are very hazardous, and may in the end do great harm. It is far better to let the obscurity remain, till time and further facts throw light upon it. I think, however, that we are too apt to lose sight of the fact, that obscurities and difficulties are exaggerated in number and importance, and that

* On the subject of the Deluge see an excellent paper, by the Rev. M. Davison, in the *Journal of the Victoria Institute*, vol. v. No. 14.

when allusions are made in Holy Scripture to natural phenomena they are often truly remarkable. All the outcry we hear, that the progress of Science is at variance with the Scriptures,—what does it come to? We might, from the things which are said, suppose that the texture of Scripture Revelation was studded with errors regarding nature, which at once disprove its Divine origin. But this is by no means the case. While the difficulties which *are* produced refer merely to the earlier chapters of the first book, the references to the phenomena of nature throughout the main body of it are not only not charged with error, but are universally admired for their true sublimity, and are as much in advance of the philosophy of even later times, as truth is in advance of error. Hear how truly we are told in the ancient book of Job, that the Almighty 'hangeth the earth upon nothing' (xxvi. 7). Hear how the prophet Amos alludes correctly to the process of evaporation from the sea, and the outpouring of the clouds so formed to fructify the earth: 'He calleth for the waters from the sea, and poureth them out on the face of the earth' (ix. 6). And Solomon too on the same subject: 'All the rivers run into the sea, and yet the sea is not full; unto the place from whence the rivers come, thither they return again' (Eccles. i. 7): that is, by evaporation and the wafting of the clouds and vapour to the hills and high-lands. No such correct natural philosophy is to be found in other ancient books. And where the phenomena of nature are made use of in a poetical way, with what sublimity do the inspired writers refer to them and bear witness to the presence of God in His own world—and that, though in poetical language, with no violation of scientific truth: 'He maketh the clouds his chariot. He walketh upon the wings of the wind' (Ps. civ. 3): and in another place; 'He made darkness His secret place; His pavilion round about Him were dark waters, and thick clouds of the skies. The Lord also thundered in the heavens, and the highest gave His voice; hail-stones and coals of fire' (Ps. xviii. 11, 13)— and it is well known that hot thunderbolts sometimes now fall from heaven and not unaccompanied with hail, regarding which the science of the present day has an interesting theory.

Take up a volume of only one hundred years old which touches on any of these matters, and there is little doubt that you will detect some gross error which the progress of human knowledge has exploded. But in the Holy Scripture we have a book, of very great antiquity, still fresh. Apparent discrepancies invariably prove the germ of new agreements. A book, so written as to touch upon many subjects of human research, and without anticipating discoveries which man can make for himself, not to contradict them when made, is certainly a paragon of wisdom and knowledge of the highest order. That the Scriptures should stand thus pre-eminent through all ages, and that they should never be behind Science however advanced, producing, that is, nothing contradictory to Science from age to age, is sufficient to convince the most sceptical of their Divine origin.

CHAPTER III.

EXAMPLES, IN WHICH SCIENCE HAS BEEN DELIVERED FROM THE FALSE CONCLUSIONS OF SOME OF ITS VOTARIES, AND THEREBY SHOWN NOT TO BE AT VARIANCE WITH SCRIPTURE, AS THEY HAVE ALLEGED.

1. FROM the great diversities which exist among the tribes of men which at present inhabit the earth it has been boldly inferred by some writers, that it is impossible that they can all have descended from common parents. The statements of Scripture, that Eve was the 'mother of all living' (Gen. iii. 20); that after the Deluge the earth was peopled by the descendants of one man, Noah (Gen. x. 32); and the declaration of St. Paul (Acts, xvii. 26) that God 'hath MADE OF ONE BLOOD all nations of men for to dwell on all the face of the earth,' are equally set aside as irreconcilable with the facts of nature. Thus the Word and Works of God have been driven once more into conflict, and upon entirely new ground.

[margin: All men of one blood.]

(1). This apparent contradiction between Revelation and Nature has been examined by the late Dr. Prichard. His facts and arguments have been collected in his invaluable work on the *Natural*

History of Man. He takes no guide but the phenomena which the various tribes present, and which well-authenticated history furnishes. And he comes to the conclusion, that there are no permanent lines of demarcation separating the several tribes or nations; that all the diversities which exist are variable, and pass into each other by insensible gradations; that there is scarcely an instance in which the actual transition cannot be proved to have taken place; and that there is everything to lead us to infer, quite irrespectively of Scripture testimony, that all the families of the earth are descended from common parents, and that at no very distant epoch.

His language is too important not to be quoted. 'The sacred Scriptures,' he says, 'whose testimony is received by all men of unclouded minds with implicit and reverential assent, declare that it pleased the Almighty Creator to make of one blood all the nations of the earth, and that all mankind are the offspring of common parents. But there are writers in the present day, who maintain that this assertion does not comprehend the uncivilised inhabitants of remote regions; and that Negroes, Hottentots, Esquimaux, and Australians, are not, in fact, men in the full sense of that term, or beings endowed with like mental faculties with ourselves. Some of these writers contend, that the races above mentioned, and other rude and barbarous tribes, are inferior in their original endowments to the human family which supplied Europe and Asia with inhabitants—that they are organically different, and can never be raised to an equality, in moral and intellectual powers, with the offspring of that race which displays, in the highest degree, all the attributes of humanity. They maintain that the ultimate lot of the ruder tribes is a state of perpetual servitude; and that if, in some instances, they should continue to repel the attempts of the civilised nations to subdue them, they will at length be rooted out and exterminated in every country on the shores of which Europeans shall have set their feet.'

Although this question is one of which the decision is not a matter of indifference either to religion or humanity, yet he follows the strict rule of scientific scrutiny which requires that we should close our eyes against all presumptive and extrinsic evidence, and abstract our minds from all considerations not derived from the matters of fact which bear immediately on the question. 'The maxim we have to follow,' he says, 'in such controversies is, *Fiat justitia, ruat cælum.* What is actually true it is always most desirable to know, whatever consequences may arise from its admission.' Taking this course, he sums up the results of his able investigation thus:—

'In the ethnographical outline which I have now concluded, the facts have been very briefly stated, and it would be difficult to recapitulate them in a shorter compass. I shall merely point out some of the most obvious inferences. The differences of men are not distinguished from each other by strongly-marked, uniform, and permanent distinctions, as are the several species belonging to any given tribe of animals. All the diversities which exist are variable, and pass into each other by insensible gradations; and there is, moreover, scarcely an instance in which the actual transition can not be proved to have taken place.'

And again, further on :—' We contemplate among all the diversified tribes who are endowed with reason and speech the same internal feelings, appetences, aversions ; the same inward convictions ; the same sentiments of subjection to invisible powers, and, more or less fully developed, of accountableness or responsibility to unseen avengers of wrong and agents of retributive justice, from whose tribunal men cannot even by death escape. We find everywhere the same susceptibility, though not always in the same degree of forwardness or ripeness of improvement, of admitting the cultivation of these universal endowments, of opening the eyes of the mind to the more clear and luminous views which Christianity unfolds, of becoming moulded to the institutions of religion and of civilised life : in a word, the same inward and mental nature is to be recognised in all the races of men. When we compare this fact with the observations which have been heretofore fully established as to the specific instincts and separate psychical endowments of all the distinct tribes of sentient beings in the universe, we are entitled to draw confidently the conclusion, that all human races are of one species and one family.'* So triumphantly is the Scripture account thus far verified by an impartial and independent appeal to facts.

It is difficult to understand how men can presume to contravene the distinct statements of Scripture on

* Prichard's *Natural History of Man*, vol. i. p. 5 ; vol. ii. pp. 644, 713, 714.

this subject on the ground that there are differences, when we compare one nation with another, which we cannot explain: for they seem to forget that there are differences in individuals, in colour and in form, who undoubtedly belong to the same nation, and regarding whose descent from common parents there is no dispute, as inexplicable as the differences between one nation and another. In the one case the common parentage cannot be denied. How, then, can it be denied in the other in the face of the Scripture statement? They reply, that in comparing nations the differences appear to be permanent, whereas the anomalies in the same nation are transitory. But this is no answer whatever to the argument; it only shows that the cases are not identical. But they are both inexplicable. Are they able to give a physiological explanation of the appearance of unusual varieties in the same nation? They confess that they cannot. Why, then, do they not deny their existence? Because they occur as matters of fact before their own eyes: and, therefore, they are compelled to resort to conjecture, and think they see an explanation in peculiarities in embryonic nutrition. But the difficulties are not really solved. They see the facts, and know, therefore, that there must be some explanation. Why not act so in the other case? What right have they gratuitously to deny the historical fact handed down in the Scriptures, that all men are of one blood, and that the nations are all descended from Noah and his sons, on the ground that they cannot yet account for difference of races? If it pleased the Almighty, either by a direct act of His own when the nations were dispersed, or by the influence of climate or other physical causes, to impress more or less permanent characteristics on different nations, is this less intelligible and more perplexing than that a man should be born with six fingers on each hand and six toes on each foot, when none of his ancestors were known to have so many, and it is quite contrary to human organization? This fact, we know, occurs occasionally in these times, as well as other anomalous varieties.* And, departures as they are from the physical constitution of man, they serve as an unanswerable rejoinder to those who object, on the ground of want of explanation as to the causes, that the historical statement of the Scriptures is not true, that all men are of one blood. We do not appeal to physiology to prove the truth of the Scripture statement; but we deny that physiology has given any proof that the Scripture statement is not true. In some paragraphs, indeed,

* Professor Huxley gives this remarkable case. 'A Maltese couple, named Kelleia, whose hands and feet were constructed upon the ordinary human model, had born to them a son, Gratio, who possessed six perfectly movable fingers on each hand, and six toes, not quite so well formed, on each foot. No cause could be assigned for the appearance of this unusual variety of the human species.' This Gratio 'married a woman with the ordinary pentadactyle extremities, and had by her four children,—Salvator, George, André, and Marie. Of these children Salvator, the eldest boy, had six fingers and six toes, like his father; the second and third, also boys, had five fingers and five toes, like their mother, though the hands and feet of George were slightly deformed; the last, a girl, had five fingers and five toes, but the thumbs were slightly deformed. The variety thus reproduced itself purely in the eldest, while the normal type reproduced itself purely in the third, and almost purely in the second and last; so that it would seem, at first, as if the normal type were more powerful than the variety.

which follow, it will be seen that the history of races does go a great way towards actually confirming, and by no means contradicting, the Scripture statement.

To the testimony of the late Dr. Prichard, I will now add that of Dr. Carpenter, equally clear and important. He concludes his examination of the subject thus :—'From the anatomical portion of our inquiry we are led to the general conclusion, *first*, that no such difference exists in the external conformation or internal structure of the different races of men as would justify the assertion of their distinct origin ; and, *secondly*, that, although the comparison of the structural characters of races does not furnish any positive evidence of their descent from a common stock, it proves that even if their common stocks were originally distinct, there could have been no essential difference between them, the descendants of any one

---

But all these children grew up, and intermarried with normal wives and husbands ; and then, note what took place : Salvator had four children, three of whom exhibited the hexadactyle members of their grandfather and father ; while the youngest had the pentadactyle limbs of the mother and grandmother ; so that here, notwithstanding a double pentadactyle dilution of the blood, the hexadactyle variety had the best of it. The same pre-potency of the variety was still more markedly exemplified in the progeny of two of the other children, Marie and George. Marie (whose thumbs only were deformed) gave birth to a boy with six toes, and three other normally formed children ; but George, who was not quite so pure a pentadactyle, begot, first, two girls, each of whom had six fingers and six toes ; then a girl with six fingers on each hand and six toes on the right foot, but only five toes on the left ; and lastly, a boy with only five fingers and toes. In these instances, therefore, the variety leaped over one generation to reproduce itself in full force in the next. Finally, the pure pentadactyle, André, was the father of many children, not one of whom departed from the normal parental type.' *Lay Sermons, Addresses, and Reviews,* pp. 291-293.

such stock being able to assume the characters of another.'* And further on :—'The general conclusion which we seem entitled to draw from the anatomical, physiological, and psychological facts to which reference has been made is, that all the human races *may have had* a common origin, since they all possess the same constant characters, and differ only in those which can be shown to vary from generation to generation.'†

The paper from which these words are quoted contains so much which bears upon my subject that I do not scruple to make large use of it, and in doing so shall adopt almost entirely the words of the author.

He remarks, that the different races of men have been divided into the *prognathous*, indicating prominence of jaws, as the Negro of Guinea Coast and the Negrito of Australia; the *pyramidal*, the Mongolian, or Tungusian of Central Asia, or Esquimaux, Types of cranial formation.

---

* *Varieties of Mankind*, an article by Dr. W. B. Carpenter, Professor of Physiology in the University of London, in the *Cyclopædia of Anatomy and Physiology*, p. 1339.

† Ibid. p. 1345. The following testimonies are also important:— 'Professor Wagner, who from his position as a teacher of comparative anatomy for many years had the fullest opportunity, as well as disposition and ability, to investigate the question of the unity of mankind. . . . in a lecture on Anthropology delivered at the first meeting of the thirty-first assembly of German naturalists and physicians at Göttingen (Sept. 1854), says, " If you ask me, on my scientific conscience, how I would formulate the final results of my investigations on this subject, I should do so in the following manner:—All races of mankind can (like the races of many domestic animals) be reduced to one original existing, but only to an ideal type, to which the Indo-European type approaches nearest" [the ideal type being, in fact, that in which Adam came forth from the hands of his Maker spotless and perfect]. To this conclusion, Weitz, to whom we owe the most elaborate accumulation of facts in Anthropology ever collected, has also arrived.' Dr. G. Moore's *First Man, and his Place in Creation*, p. 211.

or Greenlander; and the *oval or elliptic*, the natives of Western, or Southern Europe. He says, that the prognathous type, although most remarkably developed among the Negroes of the Delta of the Niger, is by no means confined to them, or to the African races in general. It is met with in various parts of the globe; and is nearly always associated with squalor and destitution, ignorance and brutality. People among whom it prevails are for the most part hunters or inhabitants of low marshy forests. In the pyramidal the most striking feature is the lateral or outward projection of the zygomotic arches. The lines bounding the face converge towards a point upwards, so as to give the skull generally a pyramidal form. The form of the face is a lozenge shape rather than oval. The greater part of the races representing the pyramidal type in a well-marked degree may be represented as pastoral nomades. But as before they may be found in remote parts of the globe, among tribes whose descent would appear to be quite different. The oval or elliptical type at once approves itself to the educated eye as distinguished by its symmetrical contour. It is found that in the skull of largest capacity amongst the races whose average is lowest, the cubical content is greater than that of the smallest skull among the highest.* With regard to

---

* See *Contributions to the Theory of Natural Selection*, by Mr. A. R. Wallace, 1870, p. 336, 338. As an example of the unfair way in which statements are sometimes made on these subjects, I give the following extract from a valuable Appendix on the Negro in Dr. Moore's *The First Man and his Place in Creation*, p. 327,—' Much stress has been laid on the situation of the opening through which the spinal cord is united with the brain. It is asserted to be further back in the Negro, and thus rather more towards its position in the Ape.

capacity I may observe, that that of the Australian savage is 82 cubic inches, and that of the Teutonic family only 94, that of the gorilla being 30; showing clearly that the savage has brain apparatus quite comparable with that of more civilized men, if he did but use it. Mr. Wallace states that 10, 26, 32, represent the average proportion of cranial capacity of apes most like man, savages, and civilized man.

Dr. Carpenter shows by examples, that these races are not always distinguishable from each other, and are connected together by a succession of gradations which renders it impossible to draw a distinct line of demarcation between them; and that they are not so invariably transmitted from generation to generation, where the purity of the race has been preserved, as to entitle them to be regarded as permanent and unalterable; but are occasionally seen to vary in a succession of generations, so that a race loses more or less completely its original type and assumes some other. He affirms that the extreme differences in the configuration of the skull, existing among the several races of men, are not greater than those which present themselves among races of domesticated mammals, *known* to have had a common origin (*e. g.* those of the hog); and are not nearly so great as those existing among other races of

It is not true if we measure the true base of the skull; it is only apparently so from the greater projection of the upper jaw in the Negro, which has no more to do with the real base of the skull than a long nose has. But the test is in the *pose* or position of the cranium on the vertebral column or backbone. The skull of the Negro is balanced on that column as perfectly as the European's. The joint surfaces are in both precisely in the centre of gravity: this is not the case in any creature but man, because he alone is formed to walk erect, and every part of his framework is constructed accordingly.'

I

mammals (as the various breeds of dog) which are generally *believed* to have had a common origin: and that, as in the case of the domesticated races, the distinctive characters are by no means clearly marked out; but that those of the typical forms are softened down in intermediate gradations, so as to present a continuous series from one type to another, in which no such hiatus is left as would justify the assumption of the specific distinctness of those types. This last fact of itself invalidates that supposition of the uniform transmission of physical characters from parent to offspring, on which the presumption of original distinctness mainly rests. For, on the theory of specific distinctness, all the descendants of the same parentage should repeat the characters of their ancestors without essential modification; whereas we find, as a matter of fact, that the distinctive characters are perpetuated in their full intensity in only a small proportion of each race, and that in great masses they are so shaded off as gradually to disappear. Ordinary observation teaches us, that not only between the parents and their offspring, but also among the different children of the same parentage, a considerable diversity of cranial formation not unfrequently exists. And on looking at the various individuals composing the ramifications of a particular family it is observable, that they agree among themselves in some peculiarity of cranial conformation which seems, from the evidence of portraits and busts, to have been transmitted downwards for centuries; and by this very character it may be separated from other families, which are in like manner distinguished for their respective peculiarities. Now there can be no reasonable doubt that many such families had originally a common ancestry. So there must have been a time when each of these peculiarities *first* manifested itself in its own branch of the common stock. For if this be not admitted we must suppose each of them to have descended from a distant pair of protoplasts. It is obvious, then, that the question of possible modification is only one of degree: and we may expect to find that even the widest diversities which have been described might have been occasioned by the sufficiently prolonged influence of external causes acting upon a succession of generations. That such has been the case to a considerable extent would appear in some instances from the direct evidence of history. In other instances it would seem a necessary inference from the facts of philology. While in others again the two classes of evidential facts, neither of them sufficent in itself, tend to confirm each other.

Dr. Carpenter brings forward the following historical illustrations. The Turks of Europe and Western Asia. These so clearly accord in physical character with the great bulk of European nations, and depart so widely from the Turks of Central Asia, that many writers have referred the former to the (so called) Caucasian rather than to the Mongolian stock. Yet historical and philological evidence sufficiently proves that the Western Turks originally belonged to the Central Asia group of nations, with which the eastern portion of their nation still remains associated, not only in its geographical position, but in its language, physical character, and habits of life; and that it is in the

<small>Historical examples of changes.</small>

western, and not the eastern, that the change has taken place. Any result arising from intermixture of the Turkish race with the inhabitants of the countries they conquered, Dr. Carpenter shows to be altogether inadequate to explain the phenomena. Another instance, he says, of the same modification is to be found in the Magyar race, which forms a large part of the population of Hungary, including the entire nobility of that country. This race, which is not inferior in mental or physical character to any in Europe, is proved by historical and philological evidence to have been a branch of the great northern Asiatic stock, which was expelled about ten centuries ago from the country it then inhabited, bordering on the Uralian mountains; and, in its turn, expelled the Sclavonian nations from the fertile parts of Hungary, which it has occupied ever since. Having thus exchanged their abode from the most rigorous climate of the old continent—a wilderness in which Ostiaks and Samoiedes pursue the chace during only the milder season—for one in the south of Europe, in fertile plains abounding in rich harvests, the Magyars gradually laid aside the rude and savage habits, which they are recorded to have brought with them, and adopted a more settled mode of life. In the course of a thousand years their type of cranial formation has been changed from the pyramidal to the elliptical; and they are become a handsome people, with fine stature and regular European features, with just enough of the Tartar cast of countenance, in some instances, to call their origin to mind. Here again it may be said, that the intermixture of the conquering with the conquered race has had a great share in bringing about this change ; but a similar reply must be returned, for the existing Magyars pride themselves greatly on the purity of their descent ; and the small infusion of Sclavonic blood, which may have taken place from time to time, is by no means sufficient to account for the complete change of type which now manifests itself. The women of pure Magyar race are said by good judges to be singularly beautiful, far surpassing either German or Sclavonian females. A similar modification, but in less degree, appears to have taken place in the Finnish tribes of Scandinavia. These may almost certainly be affirmed to have the same origin with the Lapps ; but whilst the latter retain, although inhabiting Europe, the nomadic habits of their Mongolian ancestors, the former have adopted a much more settled mode of life, and have made considerable advances in civilization, especially in Esthonia, where they assimilate with their Russian neighbours. And thus we have in the Finns, Lapps, and Magyars, three nations or tribes, of whose descent from a common stock no reasonable doubt can be entertained, and which exhibit the most marked differences in cranial characters, and also in general conformation, the Magyars being as tall and well-made as the Lapps are short and uncouth. Another instance of the same kind, which is still more remarkable if it can be entirely substantiated, is the conversion of the Georgian and Caucasian nations from the pyramidal or Mongolian to the elliptical or Indo-European type. These people are composed of an assemblage of tribes inhabiting a mountainous country, speaking languages almost unintelligible to each other, and remarkably isolated from the neighbouring nations. The beauty of form and feature and the delicacy of complexion, which characterise individuals and families among these tribes, are well known. But these attributes are for the most part confined to the families of the chiefs; and they are carefully cherished by exemption from labour, and by exclusion from undue exposure. The common people, who are engaged in the cultivation of the soil, are described by travellers as being for the most part coarse and unshapely. From a careful comparison and analysis of the languages of these races, Dr. Latham and Mr. Norris have independently arrived on different grounds, the one from the words, the other from the grammar, at the same result, viz. that they are *aptotic*, or destitute of inflexions, like the Chinese; and that the people must have been of Mongolian origin, but separated from the common stock at a very early period; the perpetuation of the very low development of their language being favoured by the peculiar character of the country in which they settled, while this tended to modify their physical conformation. For the area they occupy is at once temperate, mountainous, and wooded, the reverse of the true Mongol areas. And thus, adds Dr. Carpenter, if this view should be confirmed, we must regard the very people which has been selected as furnishing the type of the most perfect conformation, as an improved race of a decidedly inferior stock.

The Negro type is often cited as an example of the permanence of the physical character of races, and especially of types of cranial conformation. The existing Ethiopian physiognomy is said to agree with the representations transmitted to us from the remotest times in Egyptian pictures; and the physiognomy, it is further maintained, continues to be transmitted from parent to child, even where the transportation of a Negro population to a temperate climate and civilized associations, as in the United States of America, has entirely changed the external conditions of their existence. Now it is perfectly true that the Negro races which continue to inhabit their original localities, and maintain their barbarous habits of life, retain the prognathous type; and this is precisely what we should expect. But it is not true that no modification has taken place in them, either from the influence of civilization or from a change of the physical conditions of their existence, for the most elevated form of skull occurring among the African nations is found in those which have emerged in a greater or less degree from their barbarism; their civilization having been due to external influence brought to bear upon them. There is, Dr. Carpenter observes, strong evidence to show, that even the Syro-Arabian or Semitic nations may be referred to the African stock; at any rate, there are numerous tribes in the interior of Africa, whose affinity with the true Negroes cannot be disputed, and which yet present a far superior cranial organization. So that we must either regard the one form to be the result of improvement, or the other to have proceeded from degeneration. In regard to the transplanted Negroes, it is obvious that the time which has elapsed since their removal, is as yet too short to justify us in expecting any considerable alteration in cranial configuration. Many of the Negroes now living in the West Indian Islands are natives of Africa; and a large proportion of the Negro population, both there and in the United States, is removed by no more than one or two descents from their African progenitors. The climate, too, of the southern states of North America, as of the West Indies, is not very different from that of the Guinea Coast, in regard to temperature; and the low, undrained character of much of the soil which they are employed in cultivating tends to keep up the correspondence. Still, according to the concurrent testimony of disinterested observers, both in the West Indies and in the United States, an approximation in the Negro physiognomy to the European model is progressively taking place. This is particularly the case with Negroes employed as domestic servants. It is said to be frequently not at all difficult to distinguish a Negro of pure blood, belonging to the Dutch portion of the colony, from another belonging to the English settlements, by the correspondence between the features and expression of each and those which are characteristic of their respective masters. This alteration, too, is not confined to a change of form in the skull, or to a diminution in the projection of the jaw; but it is also seen in the general figure, and in the form of the soft parts, as the lips and nose. Dr. Carpenter was informed by Sir Charles Lyell, that, during a tour in America, he was assured by numerous medical men residing in the Slave states, that a gradual approximation is taking place in the configuration of the head and body of the Negroes to the European model, each successive generation exhibiting an improvement in these respects. The change is most apparent in such as are brought into closest and most habitual relation with the whites, as by domestic servitude, without any actual intermixture of races, which would be at once betrayed by the change of complexion.

Very strong evidence is furnished by philology to show that the Hottentot races are a branch of the common African stock, and the approximation of their skulls to the pyramidal type cannot be attributed to intermixture of any Mongolian race. On the other hand, among the inhabitants of Oceania, there are many races which show, more or less, the prognathous type, and this is sometimes associated with woolly or frizzled, sometimes with long and straight hair. Yet there is strong philological evidence for regarding these as descendants of colonists who spread themselves, probably by various lines of migration, from south-east Asia, and who carried to the various islands of the vast Malayo-Polynesian Archipelago the pyramidal type more or less softened down. On no other hypothesis can the extraordinary community in the fundamental elements of their languages be accounted for, the tribes which use them being in a state of complete isolation from each other. Where, as is frequently the case, the same island or group is peopled by two or more races, having different physical characters, it is always found that the greatest tendency to the prognathous type shows itself among those who appear to have longest dwelt there in a state of barbarism; and that it is most strongly marked when, to other degrading agencies, that of a low and marshy soil has been added. Even the elliptic type may occasionally present indications of degradation towards one of the others. Want, squalor, and ignorance have a special tendency to induce that diminution of the cranial portion of the skull, and that increase of the facial, which characterise the prognathous type. This cannot but be observed by any one who takes an accurate and candid survey of the condition of the most degraded part of the population of our great towns, especially the lowest classes of Irish immigrants. A certain degree of regression to the pyramidal type is also to be noticed among the nomadic tribes which are to be found in every civilized community. Among these, as has been remarked by an acute observer, 'according as they partake more or less of the purely vagabond nature, doing nothing whatsoever for their living, but moving from place to place, preying on the earnings of the more industrious portion of the community; so will all the attributes of the nomade races be found more or less marked in them; and they are all more or less distinguished for their high cheek bones and protruding jaws.'* This shows that kind of mixture of the pyramidal with the prognathous type which is to be seen among the most degraded of the Malayo-Polynesian races. Dr. Carpenter observes, that the conformation of the cranium seems to have undergone a certain amount of alteration even in the Anglo-

---

* *London Labour and London Poor*, by H. Mayhew, p. 2.

Saxon race in the United States, which assimilates it in some degree to that of the original inhabitants. Certain it is, that among New Englanders more particularly, a cast of countenance prevails, which usually renders it easy for any one familiar with it to point out an individual of that country in the midst of an assemblage of Englishmen; and, although this may chiefly depend on the conformation of the soft parts, yet there is a certain sharpness and an angularity of feature about a genuine 'Yankee,' which would probably display itself in the contour of the bones. Dr. Carpenter has observed an excess of breadth between the rami of the lower jaw, giving to the lower part of the face a peculiar squareness, which is in striking contrast to the tendency to an oval, the form most common among the inhabitants of the old country. And adds thereto, it is not a little significant that the well-marked change which has thus shown itself in the course of a very few generations should tend to assimilate the Anglo-American race to the aborigines of the country; the peculiar physiology here adverted to most assuredly presenting a transition, however slight, towards that of the North American Indians.* The evidence which I have thus gathered from Dr. Carpenter's valuable paper, shows most convincingly from history and physiology, that there is no difficulty whatever in receiving the Scripture statement that all the races of men are the offspring of one pair created at the beginning.

(2). The questioning of the descent of all the tribes of

* *Cyclopædia of Anatomy and Physiology*, pp. 1302-1345.

men from one stock has been put by some in a form for which, strange to say, the support of Scripture has been claimed. [Marginal note: Has mankind several origins?] The distinguished naturalist Agassiz, following, as it would appear, Dr. Nott of America, has avowed it as his belief, that 'there was no common central origin for man, but an indefinite number of separate creations, from which the races of men have sprung;'* and he boldly asserts that Scripture supports this view. Scripture and Science, therefore, are not at variance according to him; and we

---

* See this fully examined and refuted in Dr. Thomas Smyth's *Unity of the Human Races*, published in America, where this new and preposterous theory, while it has found some able opponents, is not wanting in warm admirers; as it appears to countenance the notion, that the slaves are of a race to whom the blessings of Christianity are not promised; for, according to this hypothesis, they are not descended from Adam!

Another writer gives this view also. '" God hath made of one blood," said the Apostle Paul, in addressing himself to the *élite* of Athens, " all nations for to dwell on the face of the earth." Such, on this special head, is the testimony of Revelation, and such is the conclusion of our highest scientific authorities. The question has indeed been raised in these latter times, whether each species of animals may not have been originally created, not by single pairs, or in single centres, but by several pairs, and in several centres, and, of course, the human species among the rest. And the *query*—for in reality it amounts to nothing more—has been favourably entertained on the other side of the Atlantic, where there are uneasy consciences, that would find comfort in the belief that Zamboo, the blackamoor who was lynched for getting tired of slavery and hard blows, was an animal in no way akin to his master. And on purely scientific grounds it is of course difficult to prove a negative in the case, just as it would be difficult to prove a negative were the question to be, whether the planet Venus was not composed of quartz-rock, or the planet Mars of old red sandstone. But the portion of the problem really solvable by science, the identity of the human race under all its conditions and in all its varieties, science *has* solved.'—Hugh Miller's *Testimony of the Rocks*, 1857, p. 249.

so far agree in our results. But both his premises being false, he furnishes an instance of an apparent agreement arising from a double error, in the interpretation both of Revelation and Nature. His first premiss is, that Science requires this view. But what is his argument? Solely one of analogy. He has started a general hypothesis, that among plants and the lower animals, identity of species does not necessarily imply identity of origin. He assumes that analogy should lead us to apply the same to the various races of men now inhabiting the world. But analogy is not demonstration. Moreover, to make his analogy worth anything even as an analogy, he must show that his theory is true in the case of *all* the lower animals, and not that it is probably the case with some. He must show that man, whom we except, is the *only* exception, before his principle of analogy can be of any service whatever. If indeed we admit this kind of reasoning, analogy will rather turn against such a conclusion. For there are varieties—individual, family, and national—in any one race of men, fully as difficult of explanation as the diversities of races one from another.* Analogy would therefore lead us to infer, that as these varieties, singular as they are, are *known* to belong to the same race, so the probability is that the several races—though differing, but not with wider distinctions than the varieties in each—all belong to one stock.

But the Professor's second premiss is, if possible, still more unwarrantable. He asserts that the Scrip-

---

* *Unity of the Human Races*, pp. 364, 371.

tures countenance this view. The groundwork of his assertion is the following passage: 'And Cain went out from the presence of the Lord, and dwelt in the land of Nod .... And he builded a city, and called the name of the city after the name of his son, Enoch' (Gen. iv. 16, 17). His inference is, that there must have been men to form the city; whereas, now that Abel was dead, Cain and his son, as far as Scripture acquaints us, were the sole descendants of Adam. The Professor thus peoples the land of Nod with descendants of another race distinct from Adam, and upon this flimsy basis grounds his assertion, even in the face of those plain and decisive statements of Scripture which I have already quoted, in which Eve is declared to be the mother of all living, and St. Paul informs us, that God made of one blood all nations of men. The abovementioned argument, strange to say, has been advanced by a writer avowing his belief in Scripture, and speaking of its sacred pages with reverence, and who holds also a high place among the scientific observers of the day. Adam was 130 years old when Seth was born, a substitute for Abel (Gen. iv. 25; v. 3). If, then, Abel, was slain in the previous year, Cain cannot have been much less then 130 years old when he went forth into the land of Nod. During this time his own descendants, according to the ordinary laws of human increase, might have amounted to a considerable population.* Cain's descendants may have been many

---

* 'An island first occupied by a few shipwrecked English in 1589, and discovered by a Dutch vessel in 1667, is said to have been found peopled after eighty years by 12,000 souls, all the descendants of four mothers.'—Quoted in Dr. Smyth's Work, p. 375.

thousands, especially when we remember the lengthened lives of those who lived before the Flood; and to these are to be added the descendants of Cain's brothers and sisters, not named (Gen. v. 4); and men enough would be found among them to build and inhabit a city. The very name, moreover, of the land to which Cain wandered, implied that it received its designation from him, and not from any people already inhabiting it; for Nod means 'wandering.'

These views and kindred views regarding the Deluge, the composition of the Pentateuch, and other such questions, Mr. Lecky informs us (*Rationalism in Europe*, vol. i. p. 323 : 3rd ed.), were advanced so far back as the middle of the seventeenth century by a French writer, M. La Peyrère, a Protestant who afterwards went over to the Church of Rome, recanted his views, and did not complete his work, the first part only having been then published.

There are two recent authors on this subject on whose works I will make a few brief remarks. They both receive the Holy Scriptures as inspired; they both suppose that there have been different races of men on the earth having different origins. Both write in a style, more flowing than solid and convincing. The first, in a book called *Pre-adamite Man*, considers that the Adam of the first chapter of Genesis was the progenitor of a race from which we are not descended; but that the Adam of the second chapter is our forefather. This, however, will not stand examination. For when our Lord said to the Pharisees, 'The Sabbath was made for man, and not man for the Sabbath. Therefore the Son of man is Lord also of the Sabbath'

(Mark, ii. 27, 28); by 'man' He must have meant the race of which the Pharisees and certainly Himself were members, and not a race from which they were not descended. But the narrative in Genesis which our Lord quotes is that of the creation of the Adam of the first chapter. Again, when our Lord is arguing with the Pharisees against divorce, He says, 'Have ye not read, that He which made them at the beginning made them male and female, and said, For this cause shall a man leave father and mother, and shall cleave to his wife; and they twain shall be one flesh?' (Matt. xix. 4, 5.) In these words our Lord quotes both the first and second chapters, and thus stamps them as relating to the creation of one and the same Adam, and not two distinct Adams.

Dr. M'Causland, in his *Adam and the Adamite*, supposes that the race of Adam is the Caucasian race only; and that at the time of Adam's creation there were already existing previously created races of men, not made as Adam was in the likeness of God, from whom the Mongol and Negro are descended. It is strange that the Almighty, who has told us of the creation of 'cattle, and creeping thing, and beast of the field after his kind,' before he created Adam, has made no mention of the creation of these human cousins. The fluent author seems strangely to have lost the meaning of St. Paul's words, that 'as in Adam all die, even so in Christ shall all be made alive' (1 Cor. xv. 22), and forgets that this *all*, dying in Adam and living in Christ, includes Greek and Jew, Barbarian, Scythian, bond and free (Col. iii. 11). This book is not wanting in fallacies in other places.

The writer says that 'the uninspired evidence of the date of man's creation, and the conditions of his early existence upon the earth, is derived from five different sources,—geology, archæology, history, language, and ethnology—whose convergent rays have lighted up the darkness of the prehistorical ages, and reveal the far-distant origin of the races of men.' For the last passage I should substitute '. . . . whose flickering tapers give light enough to show the gloom of ignorance and uncertainty which envelopes the whole subject; and to convince us that the clear light of Scripture [which he receives] is ten thousand times a better guide as to the origin and age of man's existence on the earth.'

The same author, in an interesting lecture on *Shinar* (published in 1867), propounds the same view, that there were races in the world independent of Adam, and who escaped the flood! In p. 22 he puts the entrance of the Aryans, descendants of Japhet, into India, very far too early; viz. 2000 years before Christ, in the time of Abraham. In the next page, however, he makes the date 1500, that is, 500 years later. This allows eight hundred years after the flood (even on Usher's chronology) for the first dispersion of the progenitors of all mankind, and the subsequent migration of the Aryans into India. But even this date seems too early. For the date of the first of the sacred books of the Hindoos is B.C. 1181, or thereabouts. See a note further on, under my fourth example in this chapter, drawn from Hindoo and Chinese Astronomy.

(3.) This theory, that the human race is descended

K

from several origins is broadly stated and defended in another work, published in America a few years ago, *The Types of Mankind*. The writers have entered upon a field of investigation so vast in its extent, and so full of mysteries in the physiology of organization, that no theory can possibly be evolved in our present state of knowledge, worthy of the least confidence, plausible as it may for a time appear. When all the knowledge which all the physicians have acquired has not yet led to the detection of the physiology of fever, dysentery, consumption, or cholera—diseases which are perpetually occurring under their observation, how presumptuous does it appear for any man, in the face of the statements of Revelation, to assert, that the differences of nations cannot have been brought about by natural causes. All their knowledge of physiological causation, and even action, important as it is, is but a drop in the ocean. It is too soon to frame ethnological theories. An examination of the work I have alluded to above will, I think, satisfy an impartial person that this judgment is true. The remarks I have already quoted so copiously from Dr. Carpenter are an ample answer to this book. There are, however, two parts which I will notice.

a. A large part of the volume is devoted to the accumulation of a multitude of facts regarding the races of men which have inhabited the earth during the historical period. Great research is here displayed, and much information, both valuable and interesting, is brought together; but these are so mixed up with conjectures, strong assertions, and an ill-disguised leaning towards a foregone conclusion, as to shake all confidence in the theory advanced, quite irrespectively of arguments from without, which can be brought against it. That even the alleged facts are not in every instance to be accepted, I may show by selecting an example on which a resident in India, at all acquainted with its history, can speak with some authority. 'India affords,' the writer asserts, 'a striking illustration of the fallacy of arguments drawn from climate. We there meet with people of all shades, from fair to black, who have been living together from time immemorial.' Then follow the alleged facts in illustration. 'The Rohillas, who are blonds, and situated south of the Ganges, are surrounded by the Nepauleans with black skins, the Mahrattas with yellow skins, and the Bengalees of a deep brown; and yet the Rohillas inhabit the plain, and the Nepauleans the mountains. Here we have either different races inhabiting the same climate for several thousand years without change, or the same race assuming every shade of colour. Of this dilemma, the advocates of unity may choose either horn.' But as an advocate of unity I will take neither horn; but will stab the bull by bringing forward better facts. It is well known to persons acquainted with Indian History, that the Rohillas were a colony of Afghans, who emigrated from the province of Cabul about the beginning of the eighteenth century. They are of a fairer complexion than almost any other race in Hindostan, which is quite in accordance with the elevation of the regions from which they come. With regard to the 'Nepauleans with black skins,' all that need be said is, that they are Hindoos who fled from the plains to the Nepaul hills in the fourteenth century before their Mahomedan conquerors, and settled there. The writer of the extract I have given above adds, that he might recite innumerable facts to the same effect. But his theory cannot but have a very precarious existence when such important mistakes are worked into its texture.

b. Another point I shall notice in this American work on ethnology, is the treatment which the tenth chapter of Genesis meets with at the hands of its authors. So clear and precise a document they naturally feel stands greatly in their way, as it traces the origin of many separate nations up to Noah as their progenitor. Their method of getting rid of its evidence is very simple, but not so convincing. After a lengthy and not uninteresting criticism upon the list of names, in which they aim at showing that none after those of Shem, Ham, and Japhet, are names of persons, but of nations, they endeavour to encounter the difficulty which meets them in these being all traced up to three individuals, and through them to Noah. Their own admission betrays their uneasiness. 'Our observations on these names limit themselves to guessing, as nearly as we can, what *may* have been meant by the writer of the tenth of Genesis.' And their solution shows to what an extent of licence men can go when once they abandon themselves to their own fancy, and are aiming at establishing their own preconceived conclusions. Ham, Shem, and Japhet, they say, were not men, but are merely terms symbolical of the three principal races known to the writer of the document, viz. red, yellow, and white. And Noah signifies *repose*, or *cessation;* a word which, they conceive, 'symbolized to the writer a point of time so remote from his own day, that he *ceased* to inquire further, and *reposed* from his labours in blissful ignorance, after having comprehended the vanity of human efforts to pierce that primordial gloom.'*

Men are not to be trusted, whatever their talents, or however great their knowledge, who can venture to indulge in such visionary speculations as these, in order to set aside a plain historical document.†

(4.) There is one question which has occasioned much perplexity in ethnology, viz. Whether races of men are not specially adapted to particular soils and climates to such a degree as to mark a difference of origin. Animal and vegetable tribes have their geographical and climatic limits; have not the races of men also? Europeans cannot colonize a tropical country; a tropical race is required to cultivate the soil in a tropical climate. England cannot

<small>Are races adapted to climates?</small>

---

* Compare this statement with the remarks of Mr. Rawlinson on this chapter in the note a few pages further on.

† In other parts of the volume the writers make the most they can, and much more than the case will justify, of the various readings in the Hebrew text, and pour most undeserved reproach upon the English authorized version — a matter, however, which does not affect my argument, since it is the original Scriptures alone to which I allow an appeal. Wild fancies, and unproved conjectures, such as that Ezra wrote the Pentateuch, are assumed to be true, contrary to what is obvious on the face of the narrative itself, and of the numerous seeds of internal evidence incidentally sown in the history; such as that afforded by comparing Judg. i. 21 with 2 Sam. v. 6-9, which conclusively proves that the Book of Judges, and much more the Pentateuch, was written before David's time; such as also the not unfrequent insertion of the words 'unto this day' (see, for example, Deut. iii. 14),

colonize the plains of India, nor of tropical Africa. Spain could not colonize South America, France cannot colonize Algeria. It is said that it is only constant immigration of the Anglo-Saxon race into North America, which, by keeping up the supply from the original stock, preserves the race from languishing in that country. As a general rule, certain tribes seem to be specifically adapted to certain climates and soils; and this seems to imply, they say, that different races were created for different countries.

Now, against all this it deserves to be considered, that instances are not wanting of a race becoming acclimatized to a new and even an unhealthy country, and apparently without any supply from without. This we have seen in the observations of Dr. Carpenter. Mr. Latham, quoting Mr. Hodgson, formerly of Nepal, gives an illustration :—'We are in India, and not in the best part of it. We are in a belt of forest fatal to Europeans, fatal, in many cases, to even the Hindoo of the healthier localities. Upon the extent that these

---

as if by the hand of some commentator, such as Ezra, the very insertion of which, by its abruptness, proves that the document itself bears an older date; such, again, as Exod. xvii. 14, which shows that Moses was actually commanded to commit to writing, not only the law, but historical events ; such, again, as the fact that the genealogies are given twice, both in Genesis and Ezra's own book, as I am ready to assume the First Book of Chronicles to be. Why should he have given them twice? Whereas his copying the ancient record into his own history will easily account for the repetition. The testimonies to the antiquity of the Pentateuch are indeed many, and need not be all enumerated here.

The writers endeavour to support their views by the supposed discoveries of Egyptologers, and by the flint remains now causing so much inquiry. These I pass over here, as they will be noticed further on.

unfavourable conditions affect the human frame, the evidence is conflicting. The saul forest, full of malaria everywhere, but fullest to the east of the Kossi, is endured by no human being save and except the remarkable individuals that have for ages made it their dwelling-place. Yet the Dhimal, the Bodo, and others thrive in it, love it, and leave it with regret. When others show in their fever-stricken aspects the inroads of the poison of the atmosphere, these breathe it as common air. Nay, they prefer it to the open and untainted air of the plains, where the heat gives them fever.'

In this case the people are said to be related closely in other particulars to those who surround them, and from whom they seem so much to differ in this one respect of climate: they appear not to be suspected, even by the opponents of unity, to be of a different race, and yet they have become acclimatized. And thus, while the fact remains indisputable, that some men can live and cultivate the soil, while others infallibly perish, it proves nothing as to diversity of origin. What it does prove is this, that we know little of the circumstances upon which the success or failure of acclimatization depends.* We have a further and very striking illustration of acclimatization in the admitted fact, that the Brahmin, who thrives in every part of India, is of the same stock with the Russian-Slav, who travels over North Europe in sledges, with the Scandinavian dweller on the North Cape, and with the Anglo-Saxon, who braves the cold of Labrador.

* See *British Quarterly Review*, vol. xxxi. p. 170.

(5.) Closely connected with this question is another: How far the mixture of the different nations of the human family is possible. Individuals are supposed to be of one species, and descended from the same original progenitors, when they can unite to propagate offspring, so as to perpetuate the race. If this is the case with the different nations of the earth, then all will acknowledge that they are of one original race. On this point, however, opposite opinions are expressed: and no doubt further information is required to establish this principle on ethnological grounds. Professor Huxley produces one remarkable illustration, which is in favour of the intermixing of races. 'The only trial,' he writes, 'which, by a strange chance, was kept clear of all such [extraneous] influences—the only instance in which two distinct stocks of mankind were crossed, and their progeny intermarried without any admixture from without—is the famous case of the Pitcairn Islanders, who were the progeny of Bligh's English sailors by Tahitian women. The results of this experiment, as everybody knows, are dead against those who maintain the doctrine of human hybridity, seeing that the Pitcairn Islanders, even though they necessarily contracted consanguineous marriages, throve and multiplied exceedingly.'* How this evidence can be met by the opposite side it is impossible to see. Indeed, it seems clear that not only in physical structure and mental capacity there is evidence enough, as Dr. Prichard and others have shown, to convince us that all men are of one blood; but also that the people of one part of the world can, in

Margin note: Can all races mix?

---

* *Fortnightly Review*, June 15, 1865, p. 272.

some instances, become acclimatized in another, and that races have been known to mix and produce a permanently prolific offspring—and thus the whole question is settled on merely natural principles.

But suppose that it should appear that in the present state of the world one nation cannot become acclimatized in the country of another, and that nations cannot intermix; this will furnish no ground for disputing the unity of the race. So far from it, that I think a very important argument may be drawn in favour of the truth of Scripture from this apparent contradiction. For should this be the result of careful investigation, we shall have this anomalous state of things, that on the one hand the innumerable tribes of men all the world over are so physically and mentally constituted as to give good reason to believe that they belong to the same human family descended from the same parents; and yet, on the other hand, there is some cause which prevents their permanently intermixing, or even living permanently in each other's countries. Now, as we shall have to show in our third example under this head, there is a precisely similar anomaly when the varieties of human speech are examined by philologists. They inform us that the further the inquiry has been carried, the more numerous are the indications that all languages must have been originally one; but that also evident proofs exist, that the separation into different tongues must have been by some violent and sudden cause: a discovery which precisely accords with and confirms the account in Scripture of the supernatural confusion of tongues by the direct interposition of the Almighty. Why, then, should we not believe, if in the end the facts of the population of the world as at present constituted demand it, that on that occasion of supernatural intervention some change was produced in the human frame, either by the introduction of a new element, or the suspension of one already existing, to adapt in some general way particular nations to the particular countries to which the Almighty dispersed them, when He 'determined the bounds of their habitation?'

2. Another attack has been made upon the truthfulness of the Scripture history by calling in question, <small>Differences of nations since the Flood.</small> not, as before, the physical possibility of all men being of one blood, but the sufficiency of the time, according to Scripture chronology, for bringing about the changes which are known to have existed at an early period. The objectors, under the force of evidence brought forward by Dr. Prichard and others, admit that, notwithstanding the diversities existing among the several tribes of the earth, all races may have sprung from an original stock, if we allow time enough for the operation of the causes of change. But they contend that, according to Scripture chronology, the time reckoned from the Deluge, which they take to be a new starting-point of the human family from a single pair, is altogether inadequate to the necessities of the case. It is asserted (as I have before stated) that Egyptian paintings, which may be dated at 1000 or 1500 years before the Christian era, display the forms and complexion of the Negro, the Egyptian, and some Asiatic nations distinctly marked.* The earliest of these dates coincides with the age of Moses;

---

* See Prichard's *Physical History of Man*, vol. v. p. 552.

and is, according to the Hebrew text, only 848 years, or (as will be seen under my 5th example) according to the Vatican copy of the Septuagint, 1728 years subsequent to the Deluge, when, as it is assumed, the population of the world began a second time. This interval, it is contended, even if we lengthen it by supposing the antiquity of the Egyptian monuments to have been carried too far back by some centuries, is too short for the production of such national diversities as those portrayed in Egyptian tombs.

This in itself is a simple assumption, and would present at first sight a formidable objection. But it is one, after all, which need not stagger us, nor shake our belief in the full inspiration of Holy Scripture. The instances I have quoted from Dr. Carpenter show us what great changes have been *known* to take place in periods of time of much the same length as this from Noah to Moses, even taking the shortest of the intervals mentioned above, and even in shorter periods still. Moreover, in the first place, this apparent difficulty proceeds upon the assumption, that the rate of change in man's physical condition is the same now that it was in the earlier ages of the renovated world. But it is quite conceivable that in those primitive and half-civilized times, physiological changes might take place much more rapidly than they have done more recently, and among nations of settled and civilized habits. Mr. Wallace, in his *Contributions to the Theory of Natural Selection*,* conceives, that in the infancy of mankind

---

* See the chapter on 'The Development of Human Races under the Law of Natural Selection.' I cannot quote Mr. Wallace in support of the views in the text, as he considers man to be the descendant of

what is termed natural selection would operate chiefly on the body, but in later times chiefly on the mind : and that in consequence the physical characteristics of different races got fixed very early in a permanent form. In the next place, it is assumed that the changes were always, not only slow, but gradual. It is, however, quite possible that in some cases a new type may have arisen, as it were, *per saltum*. This even Sir Charles Lyell would admit—the great advocate of the development theory, which he formerly so vigorously opposed. In his *Antiquity of Man* (p. 504), he advocates the idea of ' occasional strides ' in the process of development ; ' leaps,' which ' may have cleared at one bound the space which separated the highest stage of the unprogressive intelligence of the inferior animals from the first and lowest form of improvable reason manifested by man.' And so also does Professor Huxley. In speaking of an extraordinary birth among sheep, and another (which I shall quote further on in a note) in a Maltese family, he adds, ' In each the variety appears to have arisen in full force, and, as it were, *per saltum;* a wide and definite difference appearing at once.'* Again, ' Mr. Darwin's position might have been stronger had he not embarrassed himself with the aphorism, *Natura non facit saltum*, which turns up so often in his pages.

some inferior creature, and to have become man only when his powers were heightened by the guidance of a Supreme power in such a way as to check the influence of natural selection upon changes in his body. I make use, however, of Mr. Wallace's *principle*, that changes in the body would be more rapid when the arts of civilization were less known.

* *Lay Sermons, Addresses, and Reviews*, p. 291.

We believe that nature does make jumps now and then'—a fact, he considers, of no small importance.* As far as regards colour,† there are remarkable examples in India of the apparent want of all law. The children of the same parents—one a European, the other a Hindoo, or even an Eurasian—will be some white, and others black. Sometimes the grandchildren are dark, although the children were white, and were married to Europeans. So unaccountable are the changes in colour. In the third place, Dr. Carpenter has shown that the prognathous and other cranial forms are interchangeable, and rather indicate moral and intellectual condition than race. It is therefore gratuitous to assume that the forms seen in the paintings belonged to the nations named and now existing, and showing the same forms. The individuals drawn may have belonged to tribes by this time altogether altered. The paintings only show that there were then, as now, prognathous and other such forms among men. They teach nothing regarding the persistency of a type in a particular tribe. In the fourth place, it is a mistake to assume that the population of the world began again from a new single centre after the Deluge. Eight persons repeopled the earth. There is no evidence that Shem, Ham, and Japhet had not in them elements which made them differ widely from each other. They may have married, too, into different tribes, and their wives have been even more diversified

---

* *Lay Sermons, Addresses, and Reviews*, p. 326.

† On the subject of change of colour see some remarkable statements in the Appendix of Dr. G. Moore's *The First Man and His Place in Creation*.

than themselves. It is, then, altogether gratuitous to assert, that the races which now exist must be traced down from one man, Noah, as from a new starting-point. This at once carries our range of time 1700 years further back, to the days of Adam, or even 2309 years, according to the longer reckoning, for the operation of the causes of change; and great divergencies may have taken place in that time, as we gather from the historical illustrations from Dr. Carpenter.

3. Again, there have not been wanting men who have profanely ridiculed the account which Moses gives, not only of the origin of nations, but of the confusion of tongues. They have asserted that the variety of languages is so great, and their differences of character so wide, and history is so far from furnishing any example of the formation of even one new language, that it is inconceivable that mankind should ever have been 'of one language and of one speech' (Gen. xi. 1), and they deny too that the 'fable' of the Dispersion, as they call it, is sufficient to explain the endless and wide variations which at present prevail. But this subject has received the attention of the most learned philologists. Alexander von Humboldt, the Academy of St. Petersburg, Merion, Klaproth, and Frederic Schlegel, have all come to one conclusion by a comparison of languages, that the further philological inquiry has been carried, the more numerous are the indications that all languages must have been originally one. And in addition to this, other philologists, viz. Herder, Sharon Turner, Abel-Remusat, Niebuhr, and Balbi, have discovered evident internal proofs, that the separation

[margin: Mankind originally of one language.]

into different tongues must have been by some violent and sudden cause. So singularly do their labours confirm the literal truth of Scripture.*

This happy result is one of the triumphs of modern philology; and it is the more valuable to our argument because the scholars to whose researches it is due were not all friendly to the Mosaic account. M. Bunsen, indeed, has attempted to show from the growth of languages that a far longer interval than Scripture gives, was necessary from the Deluge downwards for the development of the languages of the world. But he altogether leaves out of his reckoning the extraordinary circumstances which immediately followed the Deluge, recorded in Scripture and attested (as we shall see) from other sources, and also borne witness to, as to the palpable effects found in the comparison of languages together, by the philologists last mentioned.

The first attempts of the defenders of the Scripture narrative during the last century failed from two causes. When languages were found to have any resemblance, one was conjectured to be the parent of the rest. But such a relationship could in no case

* See this well worked out in Wiseman's *Lectures on Revelation and Science.*

'The results of maturer and very extensive investigation prove that the 3064 languages of Adelung, and the 860 languages and 5000 dialects of Balbi, may be reduced to eleven families; and that these, again, are found to be not primitive and independent, but modifications of some original language; and that the separation between them could not have been caused by any gradual departure, or individual development, but by some violent, unusual, and active force, sufficient at once to account for the resemblances and the differences.'—*Unity of the Human Races*, p. 214.

be established. Parallel descent from a common origin seems not to have been thought of. In addition to this, analogies and resemblances were sought for in words, rather than in grammatical structure. But words do not unfrequently pass, in the lapse of time, from one phase of meaning to another in such a way as to have their original significance so overlaid and forgotten as to destroy the clue which might otherwise have been traceable. Whereas grammatical structure remains unchanged. 'The declensional and conjugational forms—the bones and sinews of a language—retain for ages both their shape and their signification with marvellous persistency.'*

A vast variety of languages, both dead and still spoken, have come under examination, and have been reduced, by attending to their grammatical structure, to a few families, the following being the chief:— (1.) the Semitic, including Hebrew, Chaldee, Syriac, Amharic, and Arabic—a family which has long been known. (2.) The Indo-European, including Sanscrit, Zend, Greek, Latin, Lithuanian, Gothic, German, and Sclavonic. The discovery of this family is of more recent date; it embraces the numerous languages spoken in modern and ancient times in the vast region stretching from Britain to Bengal. The study of Sanscrit by European scholars has led to this remarkable result.† (3.) Another family is the Scythian, of which

---

* Rev. Dr. Caldwell's *Comparative Grammar of Dravidian Languages*, p. 437.

† 'Who would have dreamed a century ago, that a language would be brought to us from the far East which should accompany *pari passu*, nay, sometimes surpass the Greek in all those perfections of form which have hitherto been considered the exclusive property of the

an interesting branch has recently been discovered latter, and be adapted throughout to adjust the perennial strife between the Greek dialects by enabling us to determine where each of them has preserved the purest and the oldest form.

'The relations of the ancient Indian languages to their European kindred are, in fact, so palpable, as to be obvious to every one who casts a glance at them, even from a distance; in part, however, so concealed, so deeply implicated in the most secret passages of the organization of the language, that we are compelled to consider every language subjected to a comparison with it, as also the language itself, from new stations of observation, and to employ the highest powers of grammatical science and method in order to recognise and illustrate the original unity of the different grammars.'—*Preface to Professor Bopp's Comparative Grammar.*

It is a fact well worthy of the student's attention, that the tenth chapter of Genesis contains the germs of this and other discoveries in ethnic as well as linguistic relationships:—

'*The Toldoth Beni Noah* [the genealogies of the descendants of Noah, in the tenth chapter of Genesis] has extorted the admiration of modern ethnologists, who continually find in it anticipations of their greatest discoveries. For instance, in the very second verse, the great discovery of Schlegel, which the word Indo-European embodies,—the affinity of the principal nations of Europe with the Aryan, or Indo-Persic stock, is sufficiently indicated by the conjunction of the Madai or Medes (whose native name was *Mada*) with Gomer or the Cymry, and Javan or the Ionians. Again, one of the most recent and unexpected results of modern linguistic inquiry, is the proof which it has furnished of an ethnic connexion between the Ethiopians or Cushites, who adjoined on Egypt, and the primitive inhabitants of Babylonia—a connexion which was positively denied by an eminent ethnologist only a few years ago, but which has now been sufficiently established from the cuneiform monuments. In the tenth of Genesis we find this truth thus briefly but clearly stated:—*And Cush begat Nimrod* . . . . *the beginning of whose kingdom was Babel.* So we have recently had it made evident from the same monuments, that *out of that land went Asshur, and builded Nineveh*—or that the Semitic Assyrians proceeded from Babylonia and founded Nineveh long after the Cushite foundation of Babylon. Again, the Hamitic descent of the early inhabitants of Canaan, which had often been called in question, has recently come to be looked upon as almost certain, apart from the evidence of Scripture; and the double mention of Sheba, both among the sons of Ham, and also among those of Shem, has been illustrated by the discovery that

called the Dravidian,* embracing the languages of

there are two races of Arabs—one (the Joktanian) Semitic, the other (the Himyaric) Cushite or Ethiopic. On the whole, the scheme of ethnic affiliation given in the tenth chapter of Genesis is pronounced "safer" to follow than any other; and the *Toldoth Beni Noah* commends itself to the ethnic inquirer as "the most authentic record that we possess for the affiliation of nations," and as a document of the very highest antiquity.'—Rawlinson's *Bampton Lectures*, 1859. Lecture II. See also *Aids to Faith*, pp. 268-271.

* The term Dravidia denotes the Tamil country, and means the country of the Dravidas; and a Dravida is defined in the Sanscrit lexicons to be 'a man of an outcast tribe, descended from a degraded *kshatriya*.' This name was doubtless applied by the Brahminical inhabitants of Northern India to the aborigines of the extreme South, prior to the introduction among them of Brahminical civilization, and is an evidence of the low estimation in which they were originally held. (See Caldwell's *Comp. Grammar*, p. 26.)

'The Dravidian languages may be regarded as most nearly allied to the Finnish or Ugrian family, with especial affinities to the Ostiak; and this supposition, which I had been led to entertain by the comparison of grammars and vocabularies alone, derives some confirmation from the fact brought to light by the Behistan Tablets [on which the political autobiography of Darius Hystaspes is recorded in old Persian, in the Babylonian, and also in the language of the Scythians of the Medo-Persian empire], that the ancient Sythic race, by which the greater part of Central Asia was peopled prior to the eruption of the Medo-Persians, belonged not to the Turkish, or to the Mongolian, but to the Ugrian stock. Taking for granted, at present, the conclusiveness of the evidence on which this hypothesis rests, the result at which we arrive is one of the most remarkable that the study of comparative philology has yet realized. How remarkable that the closest and most distinct affinities to the speech of the Dravidians of intertropical India should be those which are discovered in the languages of the Finns and Lapps of Northern Europe, and of the Ostiaks and Ugrians of Siberia! and consequently that the pre-Aryan inhabitants of the Dekhan should be proved by their language alone, in the silence of history, in the absence of all ordinary probabilities, to be allied to the tribes that appear to have overspread Europe before the arrival of the Goths and the Pelasgi, and even before the arrival of the Celts! What a confirmation of the statement that *God hath made of one blood all nations of men to dwell upon the face of the whole earth.*'— Ibid. p. 46.

South India, of which the chief are Tamil and Teloogoo. Another branch of this family is the Uralian, including the Finnish, Hungarian, and other languages. The discovery of the Dravidian branch, as distinct from the Indo-European, strikingly agrees with the ethnological difference which has long been known to exist between the inhabitants of South and North India. The migration into India in ages past of the Aryan race from the west seems to have divided in two the original inhabitants, giving the Aryan character to the mass of the population of the north of India, and leaving those of the south of the peninsula in their original Scythian, or rather Dravidian, state. To these must be added some others, such as the Mongolian, the Transgangetic, the Malay, besides the African and the American families. While the numerous languages which have been examined, and which were at one time thought to have almost nothing in common, are found to be closely allied to each other in grammatical construction, when belonging to the same family; at the same time philologists have decided, that the families have such differences as no principle of ordinary growth or expansion from a common origin can account for. Nothing but a violent change, caused by some force from without, can have created the distinct differences which now exist, if these families are the broken fragments of a once undivided whole. Thus all the results of investigation which can be considered of scientific value tend to support and illustrate the scriptural account of the original unity and miraculous confusion of languages which led to the dispersion of the descendants of Noah upon the face of the earth. What was once a formidable obstacle in our way is thus becoming more and more an unanswerable argument in favour of the harmony of Scripture and Science.

The original unity of language has received a still further illustration from the discoveries of philologists who have looked upon the science of language in the light of a physical science. Mr. Max Müller shows that an examination of the languages of the world, ancient and modern, and of the history of their changes, leads to this result, that language grows, and goes through three successive stages, which he calls the monosyllabic, the terminational, and the inflexional. In the first, of which ancient Chinese is a representative, roots are used as words, each root preserving its full independence. In the second stage, two or more roots have been united and coalesce to form a word, one at least retaining its radical independence, the last sinking down to a mere termination. This stage is best represented by the Turanian family of speech, and the languages belonging to it have generally been called *agglutinative*. In the third stage, the roots coalesce so as not to retain their substantive independence. This stage is best represented by the Aryan and Semitic families, and the languages belonging to it have sometimes been distinguished by the name of *organic* or *amalgamating*.* Language is therefore a thing which grows; and it does so by the involuntary effect of the gift of speech with which man is endowed. This represents the subject in a

---

* Mr. Max Müller's *Lectures on the Science of Languages*. First Edition, pp. 273, 274. First Series.

light which enables us to conceive an original unity in all languages: as all must have commenced with the monosyllabic, and consisted merely of roots. The impassable barriers which have hitherto separated some of the families into which languages have been divided are thus, by looking at the subject from another point of view, broken down. With regard to the origin of roots—that is, in fact, the origin of language— Mr. Max Müller rejects both the onomatopoieian and interjectional theories, and proposes the following. The 400 or 500 roots which are considered to remain as the constituent elements in the different families of language he regards as *phonetic types*, produced by a power which was inherent in human nature, and was therefore the gift of God. Man was created with the faculty of giving articulate expression to the rational conceptions of his mind. That faculty was not of his own making. It was an instinct—an instinct of the mind as irresistible as any other instinct. This creative faculty gave to each conception, as it thrilled for the first time through the brain, a phonetic expression: and this faculty would become fainter, as all other faculties do, when men ceased to want it. The fact that every word is originally a predicate, that names, though signs of individual conceptions, are all, without exception, derived from general ideas, Mr. Max Müller considers to be one of the most important discoveries in the science of language.* How remarkable do these views agree with what we read in Genesis:—'And out of the ground the Lord God formed every beast of the field and every fowl of

* *Science of Language*, pp. 369-371. First Edition.

the air; and brought them to Adam to see what he would call them: and whatsoever Adam called every living creature, that was the name thereof' (ii. 19). 'The science of language thus leads us up,' as this interesting writer says in concluding his volume, 'to that highest summit from whence we see into the very dawn of man's life on earth; and where the words which we have heard so often from the days of our childhood—"*And the whole earth was of one language and of one speech*"—assume a meaning more natural, more intelligible, more convincing, than they ever did before.'*

In addition to the argument from the philological side, there is not wanting evidence to the truth of the Scripture account of the confusion of tongues from other quarters.

There are certain traditions which linger about the very place where the event occurred. Berosus, the Chaldæan historian, a priest of Belus at Babylon, more than 2000 years ago, refers to it in a fragment which is come down to us. 'The ancient race of men,' he says, 'were so puffed up with their strength and tallness of stature, that they . . . . laboured to erect that very lofty tower which is now called Babylon, intending thereby to scale heaven . . . . the name of the ruin is still called Babel; because until this time all men had used the same speech, but now there was sent upon them a confusion of many and diverse tongues.' This is the testimony of a profane historian.

But a still more interesting confirmation has been recently discovered by M. Oppert, a very high authority

* *Science of Language*, p. 377. First Edition.

on Babylonian antiquities. He has identified the ruins of the tower of Babel with the basement of the great mound now called Birs Nimroud, the ancient Borsippa; and an inscription in the name of Nebuchadnezzar has been there discovered, in which the tower is alluded to in connexion with the confusion of tongues. The tower is called the most ancient monument of Borsippa: and these remarkable words occur in the inscription; 'A former king built it (they reckon 42 ages) but he did not complete its head: since a remote time people had abandoned it, *without order expressing their words*'*—a most distinct allusion to the confusion of tongues, which led to the dispersion, and so to the rise of many various languages.

4. At the close of the last century an attack was made† upon the Scripture account of the creation and subsequent history of man, by appealing to the astronomical works of the Hindoos, and especially to the epoch from which the calculations are ostensibly made, viz. 3102 B.C., reaching back more than 700 years before the Deluge. This epoch was the commencement of the last of the enormous periods of Hindoo chronology, called Yugas. A conjunction of the sun, moon, and planets, is spoken of in the Hindoo books as having then occurred, and is mentioned in such a manner as to imply that the fact was a matter of observation.

<small>Age of the human race according to Hindoo and Chinese astronomy.</small>

We are indebted to the late Mr. Bentley, of Calcutta, a member of the Asiatic Society, for a complete

---

* Dr. Smith's *Dict. of the Bible*—'Confusion of Tongues,' p. 1554.
† By M. Baillie: *Histoire de l'Astronomie ancienne*.

exposure of the fallacy of this objection. Indeed, the adversary's weapons are effectually turned upon himself, and one more proof is added of the harmony between Scripture and Science when rightly interpreted. By using the accurate calculations of modern astronomy it is shown, that the phenomenon of the conjunction above alluded to is a mere fable, devoid of all truth, as it could not have taken place at the date assigned, nor at any other epoch near it. It could not therefore have been, as pretended, an observed fact: but must have been determined by the Hindoo astronomers by calculating backwards, and upon imperfect data; and their vast *yugas*, or periods of time, amounting to many millions of years since the creation, are thus proved to be a pure fiction. He shows in another place, by estimating the amount of the precession of the equinoxes in the interval, that the earliest of all the observations handed down through any of the sacred books of the Hindoos— viz. the division of the zodiac into 'lunar mansions' —was in the fifteenth century before Christ.* The

* See translation of *Surya Siddhanta*, in the *Bibliotheca Indica*, published in Calcutta by the Asiatic Society, chap. i. Also, *Asiatic Researches*, vol. vi. pp. 564-572; also vol. viii. p. 208; and Bentley's *Hindu Astronomy*, pp. 1-3. Without a division of the heavens of some sort, or some fixed points to refer to, no observations of the heavenly bodies could be recorded with any approach to accuracy. This want appears to have led to the division of the zodiac into twenty-seven parts, of 13° 20' each, called Lunar Mansions, in very early times. The following are the data found in the Hindoo books, from which we can easily find the position of the equinoxes, and therefore the epoch when these mansions were formed. The longitude of star Regulus, or Cor Leonis, was 9° in Magha, the tenth mansion; and the autumnal equinox bisected the sixteenth mansion (and therefore the vernal equinox was at the beginning of the second mansion).

earliest of the Hindoo astronomical treatises extant, the *Surya Siddhanta*, bears internal marks of a much more recent origin even than this. The position of the Hindoo sphere as therein used belongs to the year A.D. 570, which is four centuries subsequent to the days of the Greek astronomer Ptolemy. There are many striking points of resemblance between the Greek system of astronomy and the Hindoo, as set forth in that treatise — especially the ingenious contrivance of representing the planetary motions by epicycles, though in a somewhat different way — which indicate very strongly a common origin, as far as these points of resemblance go. The arguments tend to show that the Hindoos borrowed from the Greeks. In the *Surya Siddhanta* (xii. 39) is mentioned a city Romaka, ninety degrees to the west — a name pointing, without doubt, to the great city, which was mistress of the Western world during the period of active commercial intercourse between India and the Mediterranean after the foundation of Alexandria, and

Hence the distance of Regulus from the vernal equinox was seven mansions and 9°, *i. e.* 102° 20'. On the 1st of January, 1850, it was 147° 45'. The difference 45° 25' shows a lapse of 3270 years, at the rate of 1° in seventy-two years, which is the precessional motion of the equinoxes. This carries us back to B.C. 1421, the fifteenth century before the Christian era.

In a similar manner the Age of the Vedas, or most ancient of the Hindoo books, is found to be a little more than 3000 years. In the *Jyotish*, or Vedic Calendar, it is stated, that when the Vedas were written the Winter Solstice was at the beginning of the twenty-third mansion, and the Summer Solstice in the middle of the ninth. The vernal equinox had made, therefore, one quarter of a mansion (or 3° 20', which is equivalent to 240 years) more regression upon the ecliptic than at the epoch above found. This will make the date of the Vedas B.C. 1181, or the 12th century before Christ.

indicating a channel through which such knowledge might be imported into India. It may be added, that there are no indications in this treatise of anything earlier than the division of the zodiac into lunar mansions, already mentioned. And therefore it may be safely said, that the notion of a greater antiquity than the Scriptures assign for the human race receives no support whatever from Hindoo astronomy.*

The so-called chronology of the Chinese has also been brought forward as an argument against the correctness of Scripture. But the early history of China is buried in obscurity. The reign of the first three sovereigns is said to have been recorded in a book, of which the Chinese of the present time avow that nothing is known. An account of the five following sovereigns is said to have been written in another book, of which a fragment relating to the joint reign of Yao and Chun, the last of the five, is placed at the head of the Shoo-king, a compilation by Confucius (or Kung-fu-tse), giving a condensed history of China from the joint reign of those two sovereigns down to his own time, in the sixth century before Christ. A

---

* In 1860 the *Surya Siddhanta* was translated from the Sanscrit into English, and published. Facilities, therefore, now exist for the study of this subject, which have been hitherto wanted. Two independent versions appeared about the same time: one procured by the writer of this from a pundit at Benares, well versed in English and Sanscrit, as well as astronomy, and published by the Asiatic Society of Calcutta; the other put forth by the American Oriental Society, and accompanied by valuable notes by Professor Whitney. That society appears to intend to publish other Indian astronomical works. If they are accompanied by notes as well executed as those which have now appeared, a most valuable addition will be made to Oriental scientific literature.

subsequent emperor, Che-whang-tse, conceived the mad scheme of destroying all the writings of the empire, under the idea of commencing a new set of annals with his own reign, that posterity might consider him as the founder of the empire. His successor, sixty years after this decree, desirous of repairing the injury, held out great rewards for the recovery of any part of the annals. The Shoo-king is said to have been thus recovered from an old man who had committed it to memory. On such an uncertain basis do early Chinese traditions rest.*

The late M. Biot attempted to fix the date of the Emperor Yao by astronomical means. He endeavoured to reconstruct the celestial sphere as he imagined it to have been at that time, and reasoned from the change in position of the vernal equinox. In this way, he fixed B.C. 2357 for the date of Yao. This is about the date of the Deluge, according to the Hebrew text. But perfect as the modern astronomical methods are for conducting such an investigation, they are altogether powerless unless the data to which they are applied are trustworthy. This is by no means the case in this instance, as I have shown elsewhere.†

5. The correctness of Scripture has been assailed from another point. Egyptian antiquities have been enlisted against Scripture history and Scripture chronology; chiefly the latter.

—deduced from Egyptian antiquities.

---

* See an account of this in *Encyclopædia Britannica*, word 'China.'

† In the *Philosophical Magazine* for January (p. 1), and June, 1862 (p. 496), I have examined M. Biot's papers in the *Journal des Savans*, of 1840 and 1859, and come to the conclusion that the results can by no means be depended upon. For the arguments I must refer my readers to the *Magazine*.

M. Bunsen, in his work on *Egypt's Place in Ancient History*, professes to have established the four following theses :—

'First: That the immigration of the Asiatic stock from Western Asia (Chaldæa) is antediluvian.

'Secondly : That the historical deluge, which took place in a considerable part of Central Asia cannot have occurred at a more recent period than the Tenth Millennium B.C.

'Thirdly : That there are strong grounds for supposing that that catastrophe did not take place at a much earlier period.

'Fourthly : That man existed on this earth about 20,000 years B.C., and that there is no valid reason for assuming a more remote beginning of our race.'*

These startling announcements are accompanied by the statement that they do not, in the opinion of the author, ' contravene in the slightest degree the statements of Scripture, though they demolish ancient and modern rabbinical assumptions.'† He adds, indeed, that ' on the contrary, they extend the antiquity of the biblical accounts, and explain for the first time their historical truth.'

The first of these alleged discoveries is directly

* Bunsen's *Egypt*, vol. iii. Preface, p. xxvii.
† I need not quote from M. Bunsen's work expressions which show the contempt with which he looks upon all who prefer adhering to Scripture chronology, to adopting his vague and baseless conclusions. But his language does not tend to assure the reader of the soundness of his cause. Indeed, the numerous admissions of difficulty which incidentally drop from his pen, would have prepared us to look for some measure of diffidence in propounding his new theory, rather than the contrary. For example, see vol. ii. pp. 124, 125, 128, 181, 184, 257, 415, 416, 428 ; vol. iii. pp. 13, 14, *et cetera.*

opposed to Scripture history, which teaches us, beyond question, that the Deluge was universal, as far as regards the human race (Gen. vi. 5-8). Whereas M. Bunsen would have us believe that the Egyptian monarchy existed before and after the Flood in an unbroken line; and therefore that this Egyptian portion of the descendants of Adam were exempted from the destruction announced in this language: 'And God saw that the wickedness of man was great in the earth . . . . and it repented God that He had made man on the earth . . . . and the Lord said, I will destroy man whom I have created from the face of the earth . . . . But Noah found grace in the eyes of the Lord' (Gen. vi. 5-8).

The other three theses are altogether at variance with the chronology, and indirectly also with the history, of the Sacred Volume.

It is true, that there is no single text in Scripture which fixes the date either of the Deluge or of the Creation of Man. Were this the case, it would be decisive. It is by a collation of passages that we gather from Scripture the periods which separate these events from each other and from the Christian era; and in this comparison there are some points of uncertainty as to the exact number of years, arising either from the various readings of the MSS., or from the manner in which some event is recorded. For example: we read of Canaan, the grandson of Noah, that he 'begat Sidon, his first-born, and Heth' (Gen. x. 15). We seem here to have the names of individual men. But, when it is immediately added that he also 'begat the Jebusite, and the Amorite, and the

Girgasite, and the Hivite, and the Arkite, and the Sinite, &c.,' the writer is evidently dealing, not with single generations, but with a condensed abstract of the origin and growth of tribes. The use here made of the term 'begat' shows us that it does not necessarily imply that the two individuals connected by this term were father and son, but may have been separated by two or more generations, the intermediate names being omitted, as probably of no sufficient notoriety. It must, however, be acknowledged that this explanation will hardly apply to the line of descent from Noah through Shem to Abram, given in Gen. xi. 10-26. The precision of the language seems to forbid our thrusting in names between those given. The ages, however, of the several patriarchs at the time of the birth of their eldest sons, are very different in the several versions of Genesis at the present day.

According to the Hebrew text, 1656 years intervened between the Creation of Adam and the Deluge; and about 1337 between the Deluge and the building of the Temple by Solomon, who, all are agreed, lived about 1000 years before the Christian era.* The

---

* The chronology noted in the margin of our English Bibles is gathered from the text as follows:—

From Gen. v. 3-28; viii. 13.

			YEARS A.M.
Seth	after Adam	.	. 130
Enos	,, Seth	.	. 105
Cainan	,, Enos	. .	. 90
Mahalaleel	,, Cainan	. .	. 70
Jared	,, Mahalaleel	. .	. 65
Enoch	,, Jared	. .	. 162

Samaritan Pentateuch, the Septuagint Version, and Josephus' History, lead to results which differ from these and from one another. The Septuagint version adds 100 years to each of the patriarchs, Adam, Seth, Enos, Cainan, Mahalaleel, and Enoch, before the birth of their sons; while it takes twenty from the age of Methuselah, and adds six to that of Lamech. The

		YEARS A.M.
Methuselah after Enoch		65
Lamech „ Methuselah		187
Noah „ Lamech		182
Deluge after Noah		600 = 1656

From Gen. xi. 10-24; xii. 4.

Arphaxad after Deluge		2
Salah „ Arphaxad		35
Eber „ Salah		30
Peleg „ Eber		34
Reu „ Peleg		30
Serug „ Reu		32
Nahor „ Serug		30
Terah „ Nahor		29
Call of Abram „ Terah		205 = 2083
Exodus „ Call of Abram (Exod. xii. 40; Gal. iii. 17; Gen. xv. 13, 16; Acts, vii. 4)		430 = 2513
Solomon's Temple after Exodus (1 Kings, vi. 1)		480 = 2993

Hence, from the Creation of Man to the Deluge is 1656 years, and from the Deluge to the Temple 1337 years. M. Bunsen makes these to be not less than 10,000 and 9000 years.

Some of the texts above quoted require a few remarks.

Exod. xii. 40. 'Now the sojourning of the children of Israel, who dwelt in Egypt, was four hundred and thirty years.' In the Samaritan Pentateuch, and in the Septuagint, the words 'who dwelt' are replaced by *and of their fathers in the land of Canaan and.* As the text stands above, it seems to say that the Israelites were in Egypt 430 years. This, as I will show, does not agree with other passages. It is quite possible, however, that those latter words may have slipt out from the Hebrew text, or that they have been inserted in the

space from the Creation to the Deluge is 2242 years according to the Vatican copy, 2262 the Alexandrine, 2265 Josephus, 1307 the Samaritan, while the Hebrew, as already stated, gives 1656. The space from the Deluge to the 70th year of Terah (Gen. xi. 26) is, according to the Vatican copy of the Septuagint, 1172 years, the Alexandrine 1072, Josephus 1002,

Samaritan text and in the LXX. to show that the expression 'sojourning' is meant to take in their dwelling as strangers in the land of promise. (See Heb. xi. 9.)

Gal. iii. 17, '. . . . the law, which was four hundred and thirty years after' [the covenant with Abraham]. This text supports the above interpretation of Exod. xii. 40. To justify, however, the precise use of it above, in order to make out the chronology of our English Bibles, the covenant must be supposed to have been made at the time that Abram was called in Haran. No doubt the promise was made there. But we know that it was repeated several times during the subsequent twenty-four years, till it was more distinctly and finally confirmed upon the promise of Isaac's birth. This doubt as to the dating of the period of 430 years is just one of those points of obscurity of no real moment, to which I allude in the text. It does not seem to have been the design to reveal to us chronological dates with minute precision. This does not justify our indulging in our own theories, irrespective of what *is* distinct in Scripture. Our uncertainty is restricted within certain defined limits.

Gen. xv. 13, 16, '. . . . thy seed shall be a stranger in the land which is not theirs, and shall serve them: and they shall afflict them four hundred years . . . . but in the fourth generation they shall come hither again.' Stephen quotes this in his address, Acts, vii. 4. In addition to the reason assigned before for the period of 430 years including the sojourning in Canaan, as well as in Egypt, the last part of this passage furnishes another argument. It was to be in the fourth generation that his seed were to return to Canaan. But 430, or even 400 years (the shorter period mentioned in this passage), is very much longer than four generations, and therefore must include something besides the bondage in Egypt, viz. the sojourning in Canaan. His prediction regarding the 'fourth generation,' was literally fulfilled. Moses and Aaron were sons of Jochabed, who was the daughter of Levi, see Num. xxvi. 59, a text which incidentally confirms the cor-the Samaritan Pentateuch 942, the Hebrew 292. The corruptions which have crept into these collateral authorities, and possibly into the Hebrew text itself, apparently have occurred since the days of Josephus. He, as well as Philo, bears strong testimony to the faithfulness of the Septuagint Translation from the Hebrew; and he states also that he translated his

rectness of our general outline. Eleazer, the priest, the son of Aaron, was, therefore, of the fourth generation from Jacob. He returned to Canaan and died there; his father, Aaron, and that generation having died in the wilderness (Josh. xxiv. 33). Could it, indeed, be made out that the whole 400 or 430 years were spent in Egypt (against which, however, the text in Numbers is conclusive), it will not support the gratuitous assumption which M. Bunsen makes to support his theory, that the Israelites must have been more than 1400 years in Egypt!

In the last text quoted, viz. 1 Kings, vi. 1, some error seems to have crept into the MSS. The period appears to have been 580 years rather than 480, judging from two other independent sources of information. From the giving of the law to the fourth year of Solomon (the year the Temple was begun, 1 Kings, vi. 1) is exactly 580 years, as determined from the following texts, Exod. xix. 1; and Num. x. 11; Josh. xiv. 10; Acts, xiii. 20, 21; 1 Kings, ii. 11. And this agrees well with the account in Judges (whereas the other will not), which makes from the death of Joshua to Samson 392 years (see from Judges, iii. 8, to xvi. 31. Prefix twenty-five years for Joshua's time in Canaan after the division of the land (Josh. xiv. 10, Judg. ii. 8); and add thirty-three years from Samson to Saul's accession (not an improbable period), and the sum is 450 years, the period mentioned by St. Paul (Acts, xiii. 20). Adding the years before the division of the land, and the reigns of Saul and of David, and four years of Solomon's, the whole equals as before, 580 years, and not 480.

These remarks show that some errors in the figures have crept into the MSS. But the truth of the history of the succession of persons and of events is not at all invalidated by them. And, therefore, though we cannot be certain about the exact chronology, we have *limits* within which the errors lie, and which forbid our stretching the periods to the undue length M. Bunsen's speculations demand of us. This is all I am contending for.

M

own Antiquities from the Hebrew Scriptures, without adding to or diminishing from the original. (See Hale's *Chronology*, vol. i. 274.) In his time, therefore, we may suppose there was no discrepancy.

But, be it observed, the discrepancies which do exist between the Hebrew text and these other authorities do not affect the historic narrative ; in the narrative they agree ; the variations are simply in the numbers of years from the addition of which the dates are found. In consequence of these difficulties, it has always been considered that Scripture chronology is invested with some degree of uncertainty as to the *exact number of years*. This does not, however, authorize our throwing aside the teachings of Scripture history. We are not at liberty to consider the Age of Man and the date of the Deluge to be so far open questions, that we may receive any theory that may be proposed. Suppose that, from some obscurity in the Scripture narrative, we could not determine the age of Isaac when Jacob was born, and were therefore unable to state the exact number of years between Abraham and Jacob; are we, therefore, to reject the history, or thrust in as many generations as we please between these two patriarchs, grandfather and grandson, because of an uncertainty regarding the interval between them ? It is upon this very principle that M. Bunsen proceeds. Although Moses and Aaron on their mother's side were grandsons of Levi (Num. xxvi. 59), M. Bunsen stretches the period of the sojourn in Egypt over fourteen centuries ! By a process of this kind, he makes the interval between Adam and the Deluge equal to 10,000 years, and from the

Deluge to the Temple to 9000; and under the cover of certain obscurities regarding the exact chronology of the Scripture narrative itself, in which there is no obscurity and no variation among the MSS. and authorities!

I have thus shown, I think, that M. Bunsen's theses are altogether at variance with Scripture. A new example is, therefore, afforded of an apparent discrepancy between Scripture and Science. I proceed to examine whether his inferences from Egyptian antiquities rest upon a better foundation; and, therefore, whether any discrepancy really does exist.

His two authorities are Manetho, an Egyptian priest, and Eratosthenes, the celebrated Greek philosopher, who had the care of the Alexandrian Library. Both of them lived in the third century before the Christian era, and more than 3000 years after the beginning of the period which they are supposed to authenticate. This precludes their being accepted as witnesses of the events they record. Moreover, their works have been lost; and fragments only, containing lists of kings, have been preserved to us by Julius Africanus, of the third century, and by Eusebius in the fourth century of the Christian era. The work of Africanus has also perished; but his list of Manetho's dynasties has been preserved by Syncellus, a Byzantine monk of the ninth century, who has also transmitted to us Eratosthenes' table of Egyptian kings.

(1.) Now, in the first place, the lists of Manetho and Eratosthenes do not agree. M. Bunsen is obliged to make various arbitrary corrections, in names even, in both of them to bring them into something like accordance. But after all his changes they differ by two centuries.*

(2.) These lists commence with the genealogy of Egyptian gods, and terminate with mortal kings. M. Bunsen assumes that the mythical age closes with the last king of the divine dynasties, and that real history begins with their successor Menes,† the first mortal king of Egypt. But this is altogether gratuitous. We should rather expect from the analogy of similar cases, that the immediate successors of the gods were not historical, and that their names were inserted to supply links between the divine rulers and the historical kings.

(3.) Then again, Eratosthenes, to whose researches M. Bunsen attaches even more importance than to those of Manetho, constructed a system of Greek, as well as of Egyptian, chronology. But his Greek system is rejected by sound scholars of the present

* 'Egyptology has a historical method of its own. It recognises none of the ordinary rules of evidence; the extent of its demands upon our credulity is almost unbounded. Even the writers on ancient Italian ethnology are modest and tame in their hypotheses compared with the Egyptologists. Under their potent logic all identity disappears; everything is subject to become anything but itself. Successive dynasties become contemporary dynasties; one king becomes another king, or several other kings, or a fraction of another king; one name becomes another name, one number becomes another number, one place becomes another place.'—Sir G. C. Lewis' *Historical Survey of the Astronomy of the Ancients*, p. 368.

† 'The similarity of the name of Menes to the names of other mythical personages—such as the Indian Menu, the Lydian Manes, the Cretan Minos, and the German Mannus—has been frequently noticed, and is certainly entitled to some weight, though it is dismissed by M. Bunsen with contempt.'—*Quarterly Review*, 1859, p. 387.

day.* Much more, then, is our confidence shaken in his Egyptian system, reaching back through so much longer a period.

(4.) Then, further, in constructing his system from the lists of Manetho and Eratosthenes, M. Bunsen makes extensive use of the Monuments and their deciphered hieroglyphics, as independent witnesses.† But there is every reason to suppose that the information of Manetho and Eratosthenes was derived largely from these very monuments. Moreover, as Mr. Grote observes, 'The monuments in themselves are no proof of the reality of the persons or the events which they are placed to commemorate, any more than the Centauro-machiæ or Amazono-machiæ on the frieze of a Grecian temple proves that there really existed Centaurs and Amazons.'‡

The case of Egypt, as here handled by M. Bunsen,

---

* Mr. Grote says, 'Both Eratosthenes and Phanias delivered positive opinions upon a point on which no sufficient evidence was accessible, and, therefore, neither the one nor the other was a guide to be followed.'—*History of Greece*, vol. ii. p. 53.

† 'Taking into consideration all the evidence respecting the buildings and great works of Egypt, extant in the time of Herodotus, we may come to the conclusion that there is no sufficient ground for placing any of them at a date anterior to the building of the Temple of Solomon, 1012 B.C.'—Sir G. C. Lewis' *Historical Survey of the Astronomy of the Ancients*, p. 440.

‡ See his *History*, vol. iii. p. 454. Also in vol. ii. p. 56, he says:—'An inscription, being nothing but a piece of writing on marble, carries evidentiary value under the same conditions as a published writing on paper. If the inscriber reports a contemporary fact which he had the means of knowing, and if there be no reason to suspect misrepresentation, we believe his assertion: if, on the other hand, he records facts belonging to a long period before his own time, his authority counts for little, except in so far as we can verify and appreciate his means of knowledge.'

is very different from those of Nineveh and Bashan, and now of Moab, in which also remains and monuments have been discovered; but in those instances they bear a most valuable testimony to the truth of Scripture history.* The Sacred Volume has handed down a continuous history, complete in itself, and independent of the information which the relics convey.

* The case of Nineveh is sufficiently notorious. I must, however, give the story about the name Belshazzar:—Daniel makes the last ruler at Babylon, when that city was taken by Cyrus, to be Belshazzar, who, he states, was slain in Babylon on that occasion. Now this name nowhere occurs in profane history. Secular writers give other names to the last king of Babylon, which can, by no artifice even, be made to coincide with Belshazzar. Nor can they be different names held by the same person; because Berosus, the Chaldean historian, states that the last king of Babylon fled to Borsippa, a place a few miles from Babylon, and was therefore absent from the city when taken by Cyrus. Here, then, appears to be a strong case against Daniel. He has been charged indeed with introducing a pure invention of his own. This apparent contradiction between sacred and profane history has been a sore thorn in our side; except with those whose confidence in the truthfulness of the Bible was such as to persuade them that some explanation must exist, though as yet not known. In 1854 a solution of this difficulty was found. In the ancient Ur, Sir H. Rawlinson discovered an inscription which solves the difficulty: for, from its words, it is gathered that Nabonadius, the last king of Babylon, associated with himself on the throne during the latter years of his reign his son Bil-Shar-Uzur, and allowed him the royal title. So that it was this son, called Belshazzar by Daniel, who was at Babylon at the time of the siege, and was there slain; while his father, who fled to Borsippa, was spared, as related by Berosus. Can any solution of a difficulty be more complete? The very stones speak out to testify to the perfect historical truthfulness of the Bible!

I cannot refrain from giving the following notes about Bashan, and then Moab:—'The results to which the researches of travellers have led are, that in the country south-east of Damascus, called the Haurân, numerous cities of great size, and in a high state of preservation, are still standing,—cities which every traveller who has seen them has felt convinced to be of a very high antiquity; they are not mere sites, in many cases not even ruins, but are still standing almost

Whereas the history of Egypt, which M. Bunsen professes to have constructed—in itself so scanty as to be but little more than a list of names—has no such claim to being independent of the monuments which are produced as witnesses.

(5.) M. Bunsen attempts to confirm his Egyptian chronology by the use of the Sothiac cycle of 1460

uninjured. The streets are perfect, the houses perfect, the walls perfect, and, what seems most astonishing, even the stone doors are still hanging on their hinges, so little impression has been made during these many centuries on the hard and durable stone of which they are built. We have described elsewhere our amazement on first beholding these massive structures, so unlike any other buildings we have ever seen or even heard of. And we could not help being impressed with the belief that, had we never known anything of the early portion of Scripture history before visiting this country, we should have been forced to the conclusion, that its original inhabitants, the people who had constructed these great cities, were not only a powerful and a mighty nation, but individuals of greater strength than ourselves. But when we consider that this Haurán is really the ancient land of Bashan, of which we are told so much in the Pentateuch, of whose inhabitants we read such marvellous things—when we recollect that when the Israelites came out of Egypt and conquered Og, the king of Bashan, it is said that he had threescore walled cities, and "all these cities were fenced with high walls, gates, and bars, besides unwalled cities a great many" (Deut. iii. 5), and that these were the cities which were built by the Rephaim in times long before Og; and, furthermore, when we find from the account in Deuteronomy that such numbers of cities are said to have existed within so small a space, that we quite marvel how the country could have been so thickly populated, yet that this same crowding together of the towns is one of the first peculiarites which we remark on visiting the Haurán at the present day; and, lastly, when we find existing among some of the towns of the Haurán the very names by which some of these old cities of Bashan were called, we cannot help being convinced that in these old cities of stone we have before us the cities of the giant Rephaim, the cities of Og, which have stood now so many centuries, and will stand as lasting monuments to all posterity of the conquest of Bashan, through the assistance given to His chosen people by the God of Israel. . . . Does it not seem as if these records of the past years—the period between successive epochs, when the heliacal* rising of the conspicuous star Sirius, or Sothis, occurs at the summer solstice, or the season when the first indications appeared of the rising of the Nile. But Professor Bockh of Berlin has shown, that it is more than probable that any coincidences which may be traced, arise from the priests having adapted their records to the cycle, which therefore offers no independent testimony. For example, the date of Menes is placed B.C. 5702, which is exactly the fourth cycle before the Christian era, astronomers having proved that Sirius did rise heliacally in Central Egypt, July

had been carefully preserved with a special design? How many cities in all parts of the world have been founded, destroyed, and founded again, and then a second time swept away, so that the very spot where they stood has long been forgotten? And might not this as well have happened in Bashan as elsewhere? Or may we not rather suppose that these cities have been suffered to remain, though for centuries hidden from the gaze of man, in anticipation of a day when men should begin to doubt the history of past times as recorded in Scripture—when doubt, growing into utter infidelity, should lead men, not only to distrust all revelation themselves, but to attempt to inoculate others with their scepticism; and then, when most required as witnesses to the truth, these old places could be again called forth to give their silent but all-convincing testimony to the accuracy of God's word?'—*Cambridge Essays*, 1858. *The Ancient Bashan and the Cities of Og.* By Cyril C. Graham, M.R.A.S., F.R.G.S., Trinity College, pp. 160-164. Inscriptions are of the time of Herod.

With regard to Moab, a most interesting discovery has been recently made of a stone on the site of the ancient Dibon, in the land of Moab, bearing an inscription in, as it would appear, the Phœnician character, and the language so closely related to Hebrew as to be scarcely distinguishable from it. It is a record by Mesha, king of Moab, with whose name the inscription begins, of his doings, and especially his war with Israel, as related in 2 Kings, iii. 4-27, and 2 Chron. xx. For the best account see *Recovery of Jerusalem*, p. 496.

* A star is said to rise heliacally when it rises at such an interval before the sun as then to be just visible.

22d, 139 A.D. Also the commencement of the reign of the gods coincides exactly with another of these cycles, the twenty-first before the Christian era.*

M. Bunsen's scheme, then, of substituting his speculations for the authentic statements of Scripture, furnishes us with one more instance of the retrograde movement which scientific research occasionally makes for a time, but which always ends—as in this instance—by adding one more proof, that no weapon formed against the sacred volume can possibly prosper.†

Before passing on to my next illustration, I propose to give some account of hieroglyphical discoveries, and to bring to view the character of the evidence which they furnish. The consequences made to flow from these supposed discoveries are so serious, that it will be well worth our while to dwell somewhat at length on the subject.

The Rosetta stone, transferred from Egypt to the British Museum in 1799, has been the key by which the mysteries of hieroglyphics upon the monuments and on the papyri are supposed *On hieroglyphic decipherment.*

---

* See Grote's *History of Greece*, vol. iii. chap. xx. Appendix. The date of Menes above given is that which Mr. Grote states as having been made out from Manetho's list. M. Bunsen, by a different interpretation, makes Manetho's computation to give 3555 years as the whole period from Menes down to Alexander; and, following Eratosthenes, he thinks this should be reduced still farther to 3284 years. This discrepancy is one of many which show how much uncertainty hangs over the whole subject.—(See Bunsen's *Egypt*, vol. iii. p. 94.)

† Dr. Hincks has shown, in an article in the *Dublin University Magazine*, for July, 1859, how utterly untrustworthy M. Bunsen is as a guide in what may be considered within the historic period, and therefore that he is 'not at all to be relied upon when he speculates.'

to have been partially unlocked. It has upon it three inscriptions, one in hieroglyphics, a second in the enchorial or demotic character of the country, and a third in Greek, which professes to give the same meaning as the other two. The hieroglyphic pictures are taken to be in some cases representations of the things of which they are pictures, or to represent one or more of the letters in the Coptic name of the thing represented. These latter are called phonetic hieroglyphics, and the former pictorial, and in some instances symbolic. Some of the hieroglyphics are called determinative, as serving the purpose of adjectives or inflections. Others are called mixed signs, or hieroglyphic groups, and are accompanied by a phonetic hieroglyphic before and one after; the purpose seeming to be this—the word signified is represented pictorially by the group, but lest it should not be rightly apprehended, the first and last letters of the word are also given in the phonetic hieroglyphics before and after it. Some of the phonetic powers of the hieroglyphics were determined from the Rosetta stone, by tracing proper names which occur in the Greek inscription in, as far as could be guessed, the corresponding places in the hieroglyphic inscription. By the exercise of the greatest ingenuity upon the monuments and papyri, the several kinds of hieroglyphics mentioned above are supposed to have been discovered. The uncertainty attending this process must be very great. (See Bunsen's *Egypt*, vol. i. p. 342.) There is nothing decisive to point out, whether any given hieroglyphic is pictorial, or phonetic, or mixed. If pictorial, there must in many instances be great doubt as to the correct idea having been caught. Thus M. Bunsen says (i. 345), 'a human figure holding its finger to its mouth represents to the Egyptian the sucking child.' Why does it not represent silence, or mystery, or contemplation? 'A female figure, forming with bent body, and head and hands hanging down, a sort of arch, represents the vault of heaven!' Then again, even should the idea of the picture be rightly caught, the word which translates it, and therefore the phonetic power of the hieroglyphic, is sought through the Coptic or old Egyptian Christian language. This is all conjecture. Moreover, the language may have greatly changed during the long period through which hieroglyphic writing prevailed, according to these writers. The ingenuity and learning which have been spent on this search into the meaning of the hieroglyphics is remarkable. But learning and ingenuity cannot compensate for the want of solid facts. So ingeniously plastic is this whole system, that no sooner does one conjecture, which has served its purpose for a time, break down, than a new and additional one is invented to meet the new difficulty, and a framework of devices —such as the pictorial, phonetic, mixed, and homophone hieroglyphics—is built up, so complicated and conjectural as almost to carry its own confutation with it. Seven hundred hieroglyphics have thus been, it is said, deciphered; many of them, in some cases as many as nineteen, representing the same sound. Here is a wide field for conjecture and mistakes. The admissions incidentally made (see, for instance, the preface to M. Bunsen's vol. iii. p. 14, and elsewhere) speak reproachfully against the value of the system, and are something like so many seeds of destruction sown in the midst of it.

On such a slender basis and by such a fragile fabric of reasoning is the system of ancient Egyptian history and chronology built up, in defiance of the plain and simple statements of Holy Scripture. It ill becomes such theorists to sneer at the adherents of the Scripture account, as the dupes of rabbinical assumptions and Jewish prejudices. It is difficult to understand how any thoughtful man can set up a comparison between the slender skeleton of names, half guessed at and half deciphered by a doubtful means of interpretation, and which Egyptologers call 'Egyptian history,' and the account handed down in the Pentateuch. Even waiving the argument of inspiration, what comparison can there be between a scanty so-called history, stretching over some thousands of years, as it is asserted, built up too much upon such a mixture of guesses and facts, and, on the other hand, a history already recorded and so handed down in MS. subject only to the vicissitudes incident to the transmission from hand to hand?

The several incidental notices of Egypt in Scripture form a consistent series of points of contact between the Hebrews and the Egyptians, which bear of themselves the stamp of authentic history. Thus, the country takes its rise and name from Mizraim (the ancient word for Egypt), grandson of Noah, born soon after the flood. (Gen. x. 6.) Abram, 500 years after this, goes down into Egypt and finds a monarch, a court, princes, and servants, and is treated in a princely manner. (Gen. xii. 10-20.) The story of Joseph, 200 years later, brings Egypt before us again, as an apparently mightier kingdom. (Gen. xxxix. 1.) After the lapse of another 200 years and more, during which the Hebrews had sojourned in Egypt, a king arose who knew not Joseph and oppressed them grievously. Egypt was a kingdom of still greater power than before, with treasure-cities, chariots and horses in abundance, soothsayers and magicians, and many signs of earthly greatness. Moses was brought up at the court, and the Lord delivered His people out of Egypt by his hand. (Ex. i.–xv.) Nearly 500 years later we find Egypt mentioned again as a great people; and King Solomon married Pharaoh's daughter. The wisdom or learning of Egypt is set forth in order to illustrate the greatness of the still greater wisdom of Solomon. (1 Kings, iii. vii. ix. xi.) Jeroboam, who lifted up his hand against Solomon, found an asylum in Egypt; and after Solomon's death, Shishak, the king of Egypt, came up against Jerusalem in the tenth century before Christ, and despoiled the Temple of its treasures, but left Rehoboam king of Judah, and returned with his acquisitions to Egypt. (2 Chron. xii.)

This event is the *earliest point of contact* between the Hebrews and Egypt which can be traced in the deciphered hieroglyphic records. Previously to this there were fifteen centuries for the development of all the greatness which made Egypt renowned, without stretching the period to several times its length, and setting up a conjectural history in the place of the distinct record in the Holy Scriptures. Even this point of contact is recognized only upon the hypothesis of the hieroglyphic decipherment being correct. In the palace at Karnak there is a representation of a Jewish figure forming part of a triumphal procession of Sheshouk (supposed to be Shishak), with a tablet on his breast and a hieroglyphic, which is read thus —'Judah King;' and supposed therefore to point to this event recorded in Scripture. (Bunsen's *Egypt*, iii. 213.) But even here there is a slight discrepancy; for the figure is bound, and forms part of a procession of triumph, apparently in Egypt. But this does not tally with the Scripture account, and indicates that the decipherment is not true and that the reference is not to this event, or that the accuracy of the Egyptian records even when made out is not to be trusted. The Scripture account need not be suspected on the ground that the Hebrews would not like to record the disaster of their king being carried away, as they have not shrunk from recording against themselves worse disasters than that.

Of the points of contact between the Hebrews and Egypt there are two, especially the first, which we might have expected would afford some information of the superior antiquity of the Egyptian kingdom, as far exceeding the limits of the Pentateuch chronology, if the Egyptians at that time really claimed it. Moses spent the first forty years of his life in the court of Pharaoh, brought up as the king's grandson, and therefore with every opportunity of becoming learned, as St. Stephen tells us he was, in all the wisdom of the Egyptians (Acts, vii. 22). And Solomon, so renowned for his erudition, was connected with the Egyptian court by marriage. Yet neither Moses nor Solomon has left any record of the superior antiquity claimed for Egypt. Moses, in particular, could not have handed down to us, as he has done, the Scripture account of man's creation and subsequent history, and the peopling of the nations, as recorded in Genesis, if the archæology claimed now for ancient Egypt were true. It is useless to attribute Moses' silence on this head to Jewish vanity and prejudice for his own nation. Surely the writers of the Old Testament are sufficiently candid in recording the weaknesses and sins of the patriarchs, and of those after them, to screen them from the charge that national vanity influenced them in the records they made.

The fact is, that the whole hieroglyphic system, the researches into which display so much perseverance, learning, and ingenuity on the part of the decipherers, is, after all, a gigantic system of conjecture, with numerous facts worked into its texture which give it the appearance of truth to sanguine minds. Science has thus been made, in the hands of able but unscrupulous men, to lift up its voice against the plain teaching of Scripture, and to set that teaching aside wherever it opposes their theory. M. Bunsen continually uses the term 'facts,' where he should have said 'inferences,' and 'discovery,' where he should have said 'hypothesis:' see, for example, vol. i. p. 441, vol. iii. p. 66, and vol. iv. p. 13. The more perseveringly this interesting inquiry into antiquities is carried on the better, if it be conducted with candour and patience, as in the end it is sure to illustrate the truth of the sacred volume.*

6. The extravagant views regarding the Age of the Human Race, advanced by M. Bunsen, have met with an apparent confirmation from an examination of the Nile-deposits which have accumulated around some of the Egyptian monuments. Excavations have been made, under the direction of Mr. Leonard Horner, in the neighbourhood of Memphis and Heliopolis, with a view to determine the thickness of the deposit made by the successive inundations of the Nile, and thence to infer the age of the country. He states as his result that the thickness of deposit down to the base of the colossal statue of Rameses II. in the area of the ancient Memphis, the date of which M. Bunsen, after Lepsius, fixes at B.C. 1631, is 9 feet 4 inches. From this he infers that the rate of deposit is about $3\frac{1}{2}$ inches in a century. He further states that it is found that the Nile-deposit goes down 30 feet lower, and that from that depth has been brought up a specimen of pottery. Assuming that the deposits have been made uniformly through all these ages, he

—from Nile-deposits.

---

* Sir G. C. Lewis, in his *Historical Survey of the Astronomy of the Ancients*—a most interesting work, which has issued from the press since the above was written—closes his examination of the hieroglyphic system of the Egyptologists, on the correctness of which he throws great doubt, with these words: 'It may be feared that the future discoveries of the Egyptologers will be attended with results as worthless and as uncertain as those who have hitherto attended their ill-requited and barren labours. The publication of an inedited Greek scholiast or grammarian might be expected to yield more fruit to literature than the vague phrases of Oriental adulation or mystical devotion, which are propounded to us as versions of hieroglyphic inscriptions.'—P. 396.

infers that man must have lived in Egypt, and have been possessed of a certain degree of civilization, at a distance of time proportionate to this depth; that is, 11,646 years before the Christian era—an epoch which altogether overleaps the boundaries of Scripture chronology. But this extraordinary conclusion will not stand before examination.

(1.) The thickness of the Nile-mud is very different in the several excavations even in the same neighbourhood, showing how irregularly the deposits were made; owing, no doubt, to local causes, which would also change from age to age.* We cannot, therefore, properly infer anything regarding the relative times which the parts above and below any fixed point, such as the base of a monument of known date, would afford; since causes may have operated, which we cannot possibly trace, to change the rapidity of deposit very materially.

(2.) Layers of sand are found in some of the excavations: and sand is more or less mixed up with the Nile-mud itself. How much at different epochs the winds from the desert may have thus added to the thickness of the annual deposit, it is impossible to conjecture. This cause, therefore, tends to throw out the calculation. Moreover, the course of the Nile has, without doubt, shifted gradually within certain limits during ages past, and has transported materials from its new channel to other parts, thus deranging all regularity in the deposits made simply by the inundations.

* The thickness of Nile-mud at spots 3100, 784, 1215 yards from the Obelisk at Heliopolis, and having different bearings from it, were 9·92, 13·25, 14·25 and 14·8, 6·67 feet. See Mr. Horner's paper in *Philosophical Transactions* for 1865, pp. 132–136.

(3.) It is hardly credible that the Nile-deposit which lies around the monument of Rameses, situated in the midst of ancient Memphis, can have been accumulating ever since the monument was erected, as this involves the highly improbable hypothesis that both the city and the temple were annually submerged many feet under water. The inundations must have been banked out. It is far more probable that the accumulations have commenced from the time when Memphis became a ruin, that is, in the fifth century of our era. This would entirely alter the rate of deposit of mud above the base of the monument, and upset the calculation altogether.

(4.) The deposits show no marks of stratification: so that the position of the specimen of pottery discovered at a depth of thirty-nine feet from the surface is not defined by anything but this depth. A thousand accidents may have brought it there, even supposing that the thirty-nine feet of earth have been accumulating undisturbed at the slow rate of three and a half inches a century—which is by no means proved. It may have been at the bottom of one of the numerous wells common in Egypt; or it may have fallen down one of the deep cracks which divide the dry soil before the inundation comes.

(5.) Mr. Horner reports that the excavators brought up from greater depths than this at which the pottery was found, and in nearly all the ninety-five excavations made near Memphis, as well as in those near Heliopolis, fragments of *burnt* brick.* This circumstance fur-

* See Mr. Horner's Papers in *Philosophical Transactions* for 1855, 1858.

nishes a very strong argument against the hypothesis of high antiquity it is supposed to support. For the brick used in Egypt in ancient times was almost invariably sun-dried, and not burnt.* The ancient temples were built of stone. It was not till the period of the Roman rule that burnt brick was commonly used. The presence, therefore, of these burnt bricks, and in so many places, militates very strongly against the theory which would give such an extravagant age to the deposits in which they are found.

(6.) A letter in the *Literary Gazette*, by W. Osburn, Esq., author of the *Monumental History of Egypt*, calls attention to another argument against Mr. Horner's calculation. He states that Abd-al-Latîf, an Arab traveller and historian, visited Memphis six centuries ago; and that he describes this statue as standing upright at that time. This altogether throws out Mr. Horner's speculations; as the upper part of the mud through which he has carried his excavations, viz. the nine feet four inches, must have taken, not 3200 years, but less than 600, to accumulate.†

---

* 'The use of the crude brick, baked in the sun, was universal in Upper and Lower Egypt, both for public and private buildings.'— Wilkinson's *Ancient Egyptians*, vol. ii. p. 26. 'I have already treated .... of the great use they made of crude brick .... those burnt in a kiln being rarely employed, except in damp stations. The southern extremity of the quay, near the temple of Luqsor at Thebes, is built of burnt brick. ... Stone was confined principally to the temples and other monuments connected with religion.'—Vol. iii. p. 316.

† I have found the following notice of a statue in a translation of Abd-al-Latif's travels. He is describing the ruins of Memphis:—
'Nous en avons mesuré une qui, sans son piédestal, avoit plus de trente coudées: sa largeur, du côté droit au côté gauche, portoit environ dix coudées; et du devant au derrière, elle étoit épaisse en proportion.

Mr. Horner's conclusions, therefore, rest upon no better basis than M. Bunsen's; and Scripture chronology and history, which their false Science has assailed, are left untouched by their ill-founded speculations, which surely are calculated to bring down contempt on all such proceedings, if not on Science itself. Other examples of miscalculations of time, arising from mistaking the age of beds, might be adduced. I will here refer to one noticed in the *Geological Quarterly Journal*, vol. 18, for 1862, pp. 218 and 450. At the former page will be found a paper which determines the age of a stratum by means of a piece of pottery to be that of the Roman occupation of Great Britain: and at the latter page a paper is given which describes a vessel found in a much older stratum as distinctly mediæval in its character! No doubt such mistakes will be made: but they show that the results of a Science still in its youth are not to be received with too much confidence.

7. The age of the human race is again brought under discussion by the disinterment of late years of the remains of man in deposits, the age of which is assumed to be far greater than Scripture chronology will allow.
—from flint and other remains of man.

When the great step was taken in establishing the vast antiquity of the earth, the fact of the non-appearance of any remains of man among the countless fossils which were disentombed and carefully studied,

Cette statue étoit d'une seule pierre de granit rouge; elle étoit recouverte d'un vernis rouge auquel son antiquité sembloit ne faire qu'adjouter une nouvelle fraicheur.'—M. Silvestre de Sacy's *Translation from the Arabic of the Account of Egypt, by Abd-al-Latif, a Physician of Bagdad*, p. 188.

was hailed as bearing strong testimony to the Scripture account of his comparatively recent appearance on the earth. Individual skeletons were occasionally discovered. But their appearance could in every instance be readily accounted for; and though an outcry was in some instances raised, that the Mosaic records were again in jeopardy, the alarm was soon allayed. But this subject has of late assumed a more serious aspect. Within the last four-and-twenty years public attention has been called to the discovery by M. Boucher de Perthes of Abbeville of a vast number of flint implements, works of human art, buried down in strata, which some assert must have been formed many ages before the date which we have been in the habit of giving to the origin of man. This discovery has been followed up by similar ones in other places. The very unexpected discoveries within the last seventeen years, announced by M. Troyon of Lausanne, of the remains of whole villages built on piles now buried low in the Lakes of Switzerland, which must have been the abode of men in a remote antiquity, and have belonged to races of which we seem to have been in entire ignorance, have, in the eyes of many, appeared to add fresh evidence, and from an entirely new source, that the supposition upon which we have been going, that no records of man are to be found in the earth, and that his recent introduction upon the theatre of the world is established by geological research, is no longer to be trusted. These new facts have awakened an amount of interest proportioned to their importance. While sceptics have triumphed in prospect of what they looked forward to as a crushing demonstration of the untruthfulness of the Mosaic records, and believers in Revelation have been unmoved in their persuasion that no discrepancy would in the end be established between Scripture and Science, there has been a large intermediate class of anxious and hesitating minds, not knowing what to think and what to believe. Considerable discussion had taken place on all hands: when the whole culminated in the appearance of a volume which had been for some time previously announced, and for which the scientific, and I may add the religious, world anxiously waited, with the highest expectations that something worthy of the author's reputation would be produced on this controverted subject. Sir Charles Lyell's past works on geology have ever been remarkable for great research and a large accumulation of facts: and this gave a guarantee that in his new work on *The Antiquity of Man* no fact would be passed by which bore on the subject, and that all would be said by him which could be said in support of the view he would adopt. His verdict is in favour of the theory, which would put back the origin of man on the earth to a date more like one thousand centuries than sixty.*

If such a result could, by any possibility, be gathered from the facts of nature, it was almost to be anticipated that Sir Charles Lyell would adopt it; for he has long been known as an advocate of the uniformitarian school of geology in opposition to the paroxysmal. Indeed his first and greatest work, on *The Principles of Geology*, was written with the avowed object of attempting

* See *Antiquity of Man*, p. 204.

to show, that, if time enough is granted, all the changes, which have occurred in the earth's strata in all time past, may have been brought about by a gradual operation of causes of no greater activity or abruptness of action than those which are now witnessed on the face of the globe. The evidence which has led so great an authority to come to a conclusion regarding the antiquity of man so opposed to the letter of Scripture, I will now examine, first giving it in abstract.

(1.) Works of Art found in Danish peat. These works have been arranged under three periods, supposed to be distinct and successive, called the Stone Age, the Bronze Age, and the Iron Age; when stone, bronze, and iron were the material of which implements were made. During these periods the Scotch-fir, the oak, and the beech respectively, are said to have flourished in Denmark in large forests. 'The minimum time,' Sir Charles says, 'for the formation of so much peat must have amounted to at least 4000 years, and there is nothing in the observed rate of the growth of peat opposed to the conclusion that the number of centuries may not have been four times as great, even though the signs of man's existence have not been traced down to the lowest stratum.'—P. 16.

(2.) Danish Shell-mounds, or kitchen Refuse-heaps containing flint knives and other instruments of stone, horn, wood, and bone, but none of bronze or iron. Oyster-shells are found in great abundance, whereas the waters where these heaps are found are now too brackish to allow oysters to live. This, however, does not necessarily bespeak a very high antiquity, as 'even in the course of the present century the salt waters have made one eruption into the Baltic. . . . It is also affirmed that other channels were open in historical times, which are now silted up.' These heaps are assigned to the earliest part of the age of stone as known in Denmark.—P. 11.

(3.) Ancient Swiss Lake-dwellings built on piles. In the shallow parts of many Swiss lakes, in from five to fifteen feet of water, ancient wooden piles are found, with their heads worn down to the surface of the mud or projecting slightly above it. These appear to have supported villages, like those in a lake in France described by Herodotus, where the Pæonians defended themselves and preserved their independence during the Persian invasion. Similar habitations are to this day found among the Papoos in New Guinea ; and I have myself seen such in the Straits of Malacca. Sir Charles assigns the most ancient of these villages to the stone age, as hundreds of implements resembling those of the Danish shell-mounds and peat-mosses have been dredged up from the mud.—P. 17. Some appear to be of later date, and are assigned to the bronze age. All of the bronze implements that have as yet been discovered are confined to Western and Central Switzerland. In the eastern lakes only those of the stone age have been found. In some of the aquatic stations, as well as in tumuli and battle-fields in Switzerland, a mixture of bronze and iron implements and works of art has been observed ; and together with them coins and medals of bronze and silver, struck at Marseilles, and of Greek manufacture, belonging to the first and pre-Roman division of the age of iron. Fifty-four species of mammalia, birds, reptiles, and fish, found in the mud, are, with one exception, still living in Europe. The exception is the wild bull, which was extant in the time of Julius Cæsar. One human skull has been dredged up, of the early stone age, and is of a type not unlike that now prevailing in Switzerland. Mr. Morlot, a Swiss geologist, assuming the Roman period to represent an antiquity of from sixteen to eighteen centuries, assigns to the bronze age a date of between 3000 and 4000 years, and to the oldest layer of the stone period an age of between 5000 and 7000 years. M. Troyon and another observer come to conclusions not differing from these. The calculations are based upon the thickness of mud deposits of regular structure in the delta of the Tinière, which falls into the lake of Geneva; and on the rate which the land has encroached on the lake of Brienne during the last 750 years.—P. 17.

(4.) Irish Lake-dwellings or Crannoges. A great number of antiquities have been discovered which are referred to the stone, bronze, or iron ages.—P. 29.

(5.) Delta and Alluvial Plain of the Nile. Here Sir Charles alludes to Mr. Horner's diggings and discovery of a specimen of pottery, which he considered to be more than 13,000 years old; a result which has met such abundant refutation, the last argument being that the pottery bears marks of being of Mahommedan date! Sir Charles gives up this high antiquity, as well he might! Mr. Horner's result was obtained by calculating the rate at which the mud is deposited by the Nile.— P. 33.

(6.) Ancient mounds of artificial construction in the valley of the Ohio, used for temples and sepulture. The age of these it is attempted to fix by the age of trees which have grown on some of these earthen works. One was cut down and found to be 800 years old. But it is thought that several generations of trees must have lived and died before the mounds were overspread with that variety of species which they supported when the white man first beheld them.—P. 39.

(7.) Mounds of Santos in Brazil. The River Santos has undermined a large mound, fourteen feet in height and about three acres in area, covered with trees, and has exposed to view many skeletons. Sir Charles had seen in 1842 fragments of the calcareous stone from this spot, containing a human skull with teeth, and in the same matrix oysters with serpulæ attached, and then concluded that the whole deposit had been beneath the waters of the sea. But he now abandons this idea, and considers that the mass was bound together by the infiltration of carbonate of lime, and that the mound may be of no higher antiquity than some of those on the Ohio.—P. 42.

(8.) Delta of the Mississippi. The lowest estimate of time required by Sir Charles for the formation of this delta is 'many tens of thousands of years, probably more than 100,000.' It is known in some parts to be several hundred feet deep. In an excavation at the depth of sixteen feet from the surface, beneath four buried forests superimposed one upon the other, charcoal and a human skeleton are said to have been found, the cranium of which is said to belong to the aboriginal type of the Red Indian race. Dr. Dowler has assigned 50,000 years as the age of this skull! but Sir Charles says he can form no opinion of the value of this result.—P. 43.

(9.) **Coral Reefs of Florida.** Professor Agassiz, assuming the rate of advance of the land to be one foot in a century, calculates that it has taken 135,000 years to form the southern half of this peninsula. In a calcareous conglomerate forming part of this, and computed, on this scale, to be about 10,000 years old, some human remains have been found.—P. 44.

(10.) **Recent Deposits of Seas and Lakes.** Under this head are given, as illustrations of change of level, the upraising of strata near Naples, within the last 250 years, containing remains of buildings; of a fresh-water deposit in Cashmere, a country subject to earthquakes and alterations in the change of level, containing remains of pottery and a splendid Hindoo temple: of the coast of Chili and Peru, where the land containing signs of man has been upheaved as much as eighty-five feet by successive shocks since the region was first peopled by the Peruvian race.—P. 45.

(11.) **Works of Art disentombed in Scotland.** Canoes have been dug up from the banks of the Clyde at an average depth of nineteen feet, and five of them lay buried in the silt under the streets of Glasgow. Nearly all of them were formed of a single oak stem, hollowed out by blunt tools, probably (Sir Charles says) by stone axes, aided by the action of fire; a few were cut beautifully smooth, evidently with metallic tools. In one of the canoes a beautifully polished celt was found, in the bottom of another a plug of cork. Sir Charles assigns different ages to these works. 'Those most roughly hewn may be relics of the stone period; those more smoothly cut, of the bronze age; and the regularly-built boat of Bankton may perhaps come within the age of iron.' Other instances are given of the discovery of remains which show that those parts of Scotland have been upheaved twenty-five feet since the Roman epoch. A rude ornament made of cannel coal has been found near Dundonald, fifty feet above the sea-level, on the surface of the boulder-clay, and covered with gravel, containing marine shells. 'If we suppose,' writes Sir Charles, 'the upward movement to have been uniform in Central Scotland before and after the Roman era, ... we should carry back the date of the ornament to fifteen centuries before our era, or to the days of Pharaoh.'—P. 47.

(12.) Coast of Cornwall. Human skulls and works of art have been found at Pertuan and Carnon, forty and fifty-three feet below the surface, the overlying strata being marine. No date is here conjectured.—P. 56.

(13.) Sweden and Norway. Assuming that a mean rate of continuous upheaval has taken place of two feet and a half a century, Sir Charles considers that the sea-coast of Norway, where post-tertiary marine strata occur 600 feet above the present sea-level, has required 24,000 years for their upheaval. Works of art and some vessels built before the introduction of iron are found sixty feet above the sea, and not higher.—P. 56.

(14.) Caves in the South of France. At Bize human bones and teeth, with fragments of pottery, were found thirty years ago, mingled with the bones of mammalia, some of extinct, others of recent species. The same was the case in a cave at Pondres near Nismes. That man and these extinct animals were contemporary was disputed, and at the time Sir Charles

Lyell himself did not agree with that view; but now, in consequence of evidence from other sources, he considers that they were contemporary, and that the antiquity of man is heightened by this argument.—P. 59.

(15.) The Caverns near Liége. More than forty of these were explored thirty years ago by Dr. Schmerling, and were found to contain human bones and bones of the cave-bear, hyæna, elephant and rhinoceros, and also flint instruments. Sir Charles agrees with Dr. Schmerling in thinking that these remains have been swept into the caves through fissures, probably by some great flood. He assigns no date to them; but says that it must depend upon the time required for certain animals to become first rare and then extinct, and also for the change to be brought about in the configuration of the Liége district, so many of these caves through which streams flowed being now laid dry and choked up.—P. 63.

(16.) Fossil Human Skeleton of the Neanderthal Cave near Dusseldorf, and Fossil Skull of the Engis Cave near Liége. The first was found in 1857, and the latter twenty-five years earlier. 'On the whole,' says Sir Charles regarding the first, 'I think it probable that this fossil may be of about the same age as those found by Dr. Schmerling in the Liége caverns: but as no other animal remains were found with it, there is no proof that it may not be newer.'—P. 78. And with regard to the two, he says, 'the two skulls . . . . have given rise to a nearly equal amount of surprise, for opposite reasons; that of Engis, because, being so unequivocally ancient, it approached so near to the highest or Caucasian type; that of the Neanderthal, because, having no such decided claims to antiquity, it departs so widely from the normal standard of humanity.'—P. 89.

(17.) Flint Implements in the valley of the Somme, in Brixham Cave, and other places. It is the discovery of these works of art by M. Boucher de Perthes at Abbeville in Picardy in 1841, and subsequently by Dr. Rigollot at St. Acheul near Amiens, which has given the impetus to this new inquiry regarding the antiquity of the human race, and has been the occasion of bringing together all the evidence which in any way bears upon the subject. There is no doubt of these flints being works of art, whatever temptation there may have arisen, since the value of these articles has become known, to impose counterfeits upon the public from avaricious motives. They are found in sand or gravel about twenty or thirty feet below the surface, and in company with fossil bones of the mammoth and other extinct species of mammals. The new attention which has been awakened to these discoveries has brought into prominence several similar ones, which were recorded at the time they were made, but had been forgotten. Thirty or forty years ago a like combination of flint implements and fossils of extinct mammals was found in a cave called Kent's Hole, at Brixham near Torquay. Three or four miles to the west, other similar caves have more recently been found and carefully examined by Mr. Prestwich and Dr. Falconer, who were deputed for this purpose by the Royal Society. So also in the valleys of the Seine (p. 150), the Oise (p. 152), the Wey (p. 161), the Thames (pp. 162, 3), the Ouse (p. 162), these implements are found, in some instances in company with the remains of extinct mammalia. In the first year of this century Mr. Frere described similar discoveries of worked flints at Hoxne near Diss, in Suffolk, so numerous that it seems to have been a manufactory of these implements; but no bones of extinct mammalia have yet been found here (p. 166). The same is the case at Icklingham in Suffolk (p. 169). In caves near Wells in Somersetshire (p. 170), and at Gower in Glamorganshire (p. 173), and near Cagliari in Sardinia (p. 177), the conjunction of human remains with those of extinct species of animals is also found. The only conclusion at which Sir Charles arrives from the various phenomena is, that man must have co-existed with certain mammalia which are now extinct: and that as it may be presumed, he thinks, that a very long period must be necessary for the dying out of these species, man must have been a sojourner on the earth far longer than has generally been supposed.

(18.) Burial Place at Aurignac in the South of France. Here human skeletons and bones of extinct animals were found associated; but the human remains having been removed and reburied in the parish cemetery, in a place which could not be recognized, before they had undergone a proper scrutiny, there is some doubt hanging over this illustration.—P. 181.

(19.) The Fossil Man of Denise in Central France. There is some doubt about its genuineness: nor, if genuine, does it appear to add anything to the argument for the high antiquity of man, for M. Pictet claims for it only the date of the last volcanic eruption of Velay.—P. 195.

(20.) Human fossil at Natchez on the Mississippi. A human bone and bones of the mastodon and megalonyx are here found in company. But considerable doubt exists whether the latter have not been washed out of a more ancient alluvial deposit.—P. 200.

I have thus laid before my readers as fair a representation as I can of the evidence Sir Charles Lyell produces, to show that man may have been an inhabitant of the earth for a period vastly longer than the Scriptures teach. Although Sir Charles nowhere himself pronounces any precise opinion regarding the age of man, he evidently approves of the deductions of others, who assign to the bronze age in Europe an antiquity of between three and four thousand years, and to the stone age before it one of between five and seven thousand, the bronze age having been followed by the iron age more than 2000 years ago; and he is ready to admit, in more places than one, that the first appearance of man may be vastly earlier than this. In speaking of the Natchez fossil, he does not call in question the possibility of its being one hundred thousand years old, and of the flint implements being still older. I have given the facts in detail, partly that it may be seen how scanty the evidence is for the end in view. No doubt these researches have brought to light interesting information, of which we were before ignorant, regarding those who have preceded us as inhabitants of this earth. Yet the general impression which we get from a survey of the evidence is its utter insufficiency to establish the conclusion drawn from it in the face of the scriptural account of man's history. The immovable conviction we have that Scripture and Science will never be proved to be at variance, goes upon the supposition, that the lines of argument shall have not a single flaw in them. No chain is stronger than its weakest link. If, then, conjecture is once only substituted for undoubted fact in the line of argument, the whole is worthless in this very grave comparison between the results of Science and the teachings of Scripture. Sir Charles Lyell's work is full of facts; but it must be said that, in reference to its subject as announced in its title, it is full of conjectures, and altogether fails to establish his position. As the human remains are found buried in peat or in deposits now upheaved, or in association with the remains of extinct species of mammalia, his argument is mixed up with four assumptions regarding, (1.) The rate at which peat grows; (2.) The rate at which river or lake deposits are made; (3.) The rate of upheaval of continents—and all these during thousands of years; and (4.) The length of time necessary for the extirpation of certain species of mammalia.

(1.) With regard to the first; the length of time he allows for the growth of the peat in Denmark is extravagantly great. Mr. Pattison, a Fellow of the Geological Society, brings forward a fact which at once shows this :—

'In the *Philosophical Transactions*, No. 330, the Earl of Cromarty records that in the west of Rossshire a considerable extent of land was, between the years 1651 and 1699 [that is, in forty-eight years], changed from a forest into a peat moss, from which turf was cut.'* He says also, that 'the frequent discoveries of mediæval objects low down in fen deposits, and the experience of all those who have had to do with the management of peat land, lead to the conclusion that 2000 years constitute ample allowance for the growth of all the peat on the present surface of the globe.' And he quotes Sir Charles' own book, *The Principles of Geology*, against him :—' In treating, in that work, of the recent origin of peat mosses, he quotes the case of Hatfield moss in Yorkshire, "which appears clearly to have been a forest 1800 years ago ;" and after giving other instances, states that "a considerable portion of the peat in European peat-bogs is evidently not more ancient than the age of Julius Cæsar," and, what is most material to the present inquiry, quotes from Gerard, the historian of the valley of the Somme, a statement, that " in the *lowest* tier of that moss was found a boat loaded with bricks."' Moreover, Sir Charles in the present work states, that the growth of peat varies with the humidity of the climate, with the heat and cold, as well as from diversity in the species of plants (p. 111). To pronounce, therefore, that the Danish peat must have required 4000 years, and may have taken four times as long, and to build such a momentous result upon this estimate, when circumstances may have so varied in a much shorter interval of time as most materially to alter the rate, is an assumption which can by no means be regarded as trustworthy. On this, in part, hangs the estimate of the dates of the Stone and Bronze Ages.

* *The Antiquity of Man ; an Examination of Sir Charles Lyell's recent work.* By S. R. Pattison, F.G.S.

The existence of fossil pines low down in the peat, and in the layer above oak-trees, neither of which flourishes now in Denmark, both are superseded by the beech, Sir Charles considers to be a strong evidence of the great age of the peat, especially as he gathers from Cæsar's *Commentaries* that the trees did not grow there in his time, but the beech, as at present. This, however, Mr. Pattison calls in question, by stating that Cæsar has asserted that the fir did not grow in Britain, whereas it is known that fir was actually used in Britain by the Romans. If this can be established it would turn the argument the other way, and show that the whole peat is the growth of only about 2000 years. On such slender and uncertain grounds do some of these sweeping conclusions rest! Mr. Pattison adds, that 'the superposition of oak timber in the bogs is easily accounted for without calling in the aid of thousands of years. The process and its progress are matters of ordinary observation. A clump of pine-trees grows, with here and there an oak; the firs are the first to become old and feeble; some of them fall and begin to decay; the tiny streamlet meandering through the wood is dammed up; mosses grow; the firs all fall; the bog increases; the more hardy oak yields next; the bog still increases; the birch and alder survive on the driest spots; but these, too, are ultimately engulphed.'

The far-distant epoch which is assigned to the Stone Age, the most ancient of the three ages, is brought down within the probabilities of our current chronology, by the fact, made known to us by a writer in *The Times*, that this stone age still continues in the neighbourhood of Cape Horn and other parts of the world.*

Even the archæologist Worsaae, who has done more than any other individual in opening this vast field of inquiry into the stone, bronze, and iron ages,

---

* 'Tierra del Fuego, with its innumerable islands and rocky islets, like the mountain ranges half sunk in ocean, combines every variety of aspect — storm-beaten rocky summits, several thousand feet above the sea, glaciers so extensive that the eye cannot trace their limits, densely wooded hill-sides, grand cascades and sheltered sandy coves — altogether such a combination of Swiss, Norwegian, and Greenland scenery as can hardly be realized or believed to exist near Cape Horn. Yet even there — by lake-like waters, though so near the wildest of oceans — thousands of savages exist, and migrate in bark canoes!

'In 1830 four of those aborigines were brought to England. In 1833 three of them were restored to their native places (one having died). They had then acquired enough of our language to talk about common things. From their information and our own sight are the following facts:—The natives of Tierra del Fuego use stone tools, flint knives, arrow and spear-heads of flint or volcanic glass, for cutting bark for canoes, flesh, blubber, sinews, and spears, knocking shell-fish off rocks, breaking large shells, killing guanacoes in time of deep snow, and for weapons. In every sheltered cove, where wigwams are placed, heaps of refuse—shells and stones, offal and bones—are invariably found. Often they appear very old, being deeply covered with wind-driven sand or water-washed soil, on which there is a growth of vegetation. These are like the "kitchen middens" of the so-called "stone age" in Scandinavia.

'No human bones would be found in them (unless dogs had dragged some there), because the dead bodies are sunk in deep water with large stones, or burnt. These heaps are from six feet to ten feet high, and from ten or twenty to more than fifty yards in length. All savages in the present day use stone tools, not only in del Fuego, but in Australia, Polynesia, Northernmost America, and Arctic Asia. In any former ages of the world, wherever savages spread, as radiating from some centre, similar habits and means of existence must have been prevalent; therefore casual discovery of such traces of human migration, buried in or under masses of water-moved detritus, may seem scarcely sufficient to define a so-called "stone age."'—April 28 [1863]. R. Fitzroy.

requires, according to Mr. Pattison, a far less remote antiquity for them than Sir Charles does. Worsaae's conclusion is—

'That the three periods had not long intervals of time between them, but that implements of stone continued to be the tools of the poorer folk, whilst the more advanced families indulged in weapons, first of bronze, and afterwards of iron. Worsaae concludes his review of the chronology of these dim and mysterious ages by saying:—"It will be at once seen that the stone period must be of extraordinary antiquity; if the Celts possessed settled abodes in the west of Europe more than 2000 years ago, how much more ancient must be the population which precedes the arrival of the Celts! A great number of years must pass away before a people like the Celt could spread themselves over the west of Europe, and render the land productive; it is therefore no exaggeration if we attribute to the stone period an antiquity of at least 3000 years. There are also geological reasons for believing that the bronze period must have prevailed in Denmark five or six hundred years before Christ."'

Sir Charles Lyell thinks that 'the slowness of the progress of the arts of savage life is manifested by the fact, that the earlier instruments of bronze were modelled on the exact plan of the stone tools of the preceding age, although such shapes would never have been chosen, had metals been known from the first' (p. 377). Why should not this argument be used to show that the stone and bronze periods were nearer each other than Sir Charles supposes, and even contemporary, the 'poorer folk' using the commoner articles?* In order to push back the stone age as far as possible he says, 'We see in our own times, that the rate of progress in the arts and sciences proceeds in a geometrical ratio as knowledge increases; and so when we carry our retrospect into the past, we must be prepared to find the signs of retardation augmenting in a like geometrical ratio: so that the progress of a thousand years, at a remote period, may correspond to that of a century in modern times, and in ages still more remote, man would more and more resemble the brutes in that attribute which causes one generation exactly to imitate in all its ways the generation which preceded it' (p. 377). Is not this special pleading? Has the writer forgotten the civilized state of ancient Egypt, the state of cultivation of ancient

* This Sir J. Lubbock allows in his *Pre-Historic Time*, p. 62, first edition; p. 76, second edition.

The *Quarterly Review* draws an argument from the architecture of India, which I can fully confirm, to show how unsafe it is to infer from the co-existence in the same place of stone and bronze implements, that the people who used the stone implements must have preceded those who used the bronze ones. The Hindoos never use the arch in their temples or monumental buildings, though the Mahomedans in India have made constant use of it for several centuries. The writer gives an instance: 'There exists in Guzerat the city of Ahmedabad, built by the Moslems between the years A.D. 1411 and 1583. It contains numberless mosques, palaces, and public buildings, not one of which is without arches. In the same province there is the holy city of Palitana, containing numberless temples of the Jains and Hindus. Many of these were erected during the last fifty years, most of them either built or re-built since Ahmedabad ceased to be a capital; but, so far as is known, not one arch exists within its walls.' The writer truly says, 'Any antiquary fresh from Europe, if he saw two buildings standing side by side, one of which was arched throughout, the other exhibiting the most extraordinary "tours de force" to support its roof and cover its openings by stone beams, would at once decide that the latter must be the older, because it must have been erected before the invention of the arch.'
—*Quarterly Review*, April 1870, p. 438.

Greece in the fine arts and philosophy; and has he compared them with those countries now? Have arts once practised never been lost, even in civilized countries? Is it impossible even to reverse the view, and to conceive that the stone age might be even more modern than the bronze, as the people using the bronze implements might have been driven to the necessity, from various causes, and among them the failure of metal, to be satisfied with stone ones?*

* In an article on this subject in *Macmillan's Magazine*, April 1863, the parallelism, or rather contrast, between these ages and the golden, silver, brazen, and iron periods of Hesiod, is pointed out. This contrast 'embodies the distinction between what we may call the theological and the scientific view of man. The one sees him fall from an original state of perfection; the other as slowly and painfully working his way upward to that state. This contrast is the more interesting from the fact of the theologic point of view having been taken by a pagan writer —showing how deep in the human heart is rooted this conception of a primeval state of intelligent innocence.'

The following extracts from the *Saturday Review*, August 12, 1865, are worthy of note:—

'The evidence which has been accumulated recently respecting the antiquity of the human race and its earliest conditions is in the highest degree interesting, but it is still in too obscure and unconnected a state to support or indicate conclusions which can be said to be in any way satisfactory. . . . When we attempt to employ them for the purposes of a theory, we must be very sanguine or very hasty if we think that they will yet bear very much being put upon them. . . . It is convenient, and perhaps inevitable, to throw knowledge which occupies our thoughts into the shape of a theory; but it is important to remember that we do not yet really know enough to generalize with anything like good grounds about "ages" of stone and metals, and to conclude that they must have actually succeeded one another in time, in an order of diminishing rudeness. . . . The utmost that these remains enable us to do is to conclude something of certain races in a corner of the world, probably, at any rate possibly, driven into it from earlier seats; they contribute but little light to the larger and more interesting questions connected with the early condition and progress of mankind. And these remains themselves are for the present hopelessly isolated. All existing collec-

(2.) The date of the Stone and Bronze Ages is guessed at also from the rate of deposit of mud in the delta of the Tinière, as already stated. But here, again, to base such a conclusion upon an assumption that the deposit has proceeded uniformly through so many ages, or that the rate at which land has encroached on a lake has been uniform, so as to form a fair scale for calculation, is altogether gratuitous. Many circumstances in the past, of which we are now

tions, numerous and abundant as they are, fail to supply a thread which connects one group with another, either in the line of descent or in collateral relationship. We cannot find the clue to pass from stone to bronze, or from bronze to iron. Further, it is very precarious to make rudeness in workmanship, or difference in material, a test of relative antiquity. It appears to be beyond a doubt that certain communities —those, for instance, of the Danish shell-mounds, and of certain of the Swiss lake-villages—used only stone, and not metal. It is probable in a high degree that stone in general, as a material, preceded metal. But beyond this all seems still open to question. It seems perfectly conceivable that some tribes may have gone on using their rude chipped flints long after others were using polished stone, long after others were using bronze and iron. It is perfectly conceivable, because tribes are found doing so at this moment. Much must have depended, especially when intercourse was so restricted, on the presence of a certain material, and on the scarcity or plenty of it, and also on the social condition and habits of the race. Men often change with extraordinary rapidity when once a great improvement is opened to them; but they also often cling to what they are accustomed to, even in the presence of what is in itself more convenient; for mere custom by itself makes a great part of convenience. We know that great varieties of civilization may long subsist together— if divided by sufficient barriers of blood, of religion, of interests, of traditional habits—not only contemporaneously, but in close neighbourhood; and it is not more safe to conclude that villages in the Swiss lakes which used bronze must have been more recent than neighbouring ones which used stone, than it would be to conclude, from traces of social life, that the English of the Pale and the Irish outside it could not have been contemporaries in the sixteenth century. Again, the relation, in point of time, of bronze to iron is far too uncertain to warrant us in making an age of iron after an age of bronze. It may be probable that in certain

utterly ignorant, may have occurred to throw out such calculation altogether. Sir Charles quotes a passage from Mr. Geikie which abundantly illustrates this (see page 50). It shows to what very great irregularities the deposit of silt is subject, owing to change of currents, arising from various circumstances. Surely the signal failure of the argument from Nile-deposits, and the escape we have had from a gross imposition in chronology, based apparently upon physical phenomena,

races bronze was used before iron, in preference to it, or at any rate instead of it; but, as a general rule, we can but guess, and our grounds for guessing are not very good. We are in absolute ignorance of everything connected with the first use of the metals; how and where they were applied to the purposes of daily life; under what circumstances of discovery, or foreign introduction and teaching, they came to be employed in Europe. And whatever can be said of the difficulty of reducing and working iron may be balanced by the difficulties, at least equal, on the other side, connected with the materials and the invention of bronze. Bronze presupposes two remarkable things: the supply of a rare metal—tin; and the knowledge how most effectively to use it in an alloy with copper. We at present only know of one adequate source of tin in Europe—namely, Cornwall; and if that was the source before the days of the Homeric poems, a commerce among nations is implied, and with this commerce a civilization, which take away all improbabilities connected with the full use of iron. People who could have invented bronze, and procured the materials for it, could surely have procured and wrought iron; and something else besides rudeness and chronological priority must be supposed to account for the preference of one metal to the other. . . . In the earlier conditions of the world, when there was apparently much migration under very different conditions from those under which migration went on later and goes on at present, it is conceivable that vanquished and dispossessed races, driven into ruder and poorer seats, cut off from the facilities of life which they once had and which commerce did not supply, and reduced to coarser and scantier appliances, may have lost arts instead of gained them in the progress of time. . . . It is clear that in the undeveloped state of the world, when life was hard, and communication and trade so difficult, the possibility must be fully taken into account of men going back under unfavourable conditions and giving up metal from the

should teach us to beware of such specious reasonings. These things bring down discredit upon geology, or rather upon the speculations of some who indulge their fancies too freely.*

We can fix only the *relative* order of changes in the earth's strata with any certainty. Any approximation to the actual number of years which any change has occupied, is attained by assuming the rate of change, and fixing upon some date within historical times

sheer impossibility of getting it. . . . It seems a matter of regret that this subject should be mainly studied with reference to the mere question of the antiquity of the race. Our knowledge is in much too crude and disjointed a state to help us much in this particular point, which appears to derive an importance—an undue importance, as we think—from considerations indirectly connected with it. Where, as in this case, so little explains itself, and so much has to be learnt before we can be said to know anything, it would be more fruitful to study these singular remains for themselves, and to leave aside the question of age, or even of the relation, in point of time and succession, of one group to another, till we have better means of determining it than we possess. There is more likelihood of their falling at last into their proper place, if, in the first instance, they are investigated simply for their own sake, as fresh evidences of unknown states of human society, without any thought of proving either their antiquity or their lateness. They would not be a bit less curious, so far as we could find out anything about them, though we looked at them without reference to the controversies with which they have been connected, and though we doubted whether they came before civilisation in the East, or long after it had spread westwards, or whether there were grounds for assigning them a date at all.'

* The following extract from an American periodical forwarded to the *Record*, August 4, 1865, throws damaging doubt upon some other similar measures of time :—

'These, however, are small matters compared with Lyell's enormous blunders in the computation of our American processes of construction of continents. He goes to work ciphering out the age of the delta of the Mississippi, and declares the Mississippi was at least one hundred thousand years making the delta, and makes a wonderful display of figures to prove it. The United States' Government has set

to form the basis or starting-point of the calculation. In both these we are liable to error. Some years ago, a geological lecturer of no ordinary note asserted that the volcanoes of Auvergne, in Central France, have not been in activity for many ages—certainly not since the days of Julius Cæsar, who pitched his camp there in perfect safety; and he took the intervening period of nearly 2000 years as the first step for measuring the antiquity of the deposits in those parts. Whereas, ten

its surveyors, Abbot and Humphrey, to work with chain and Jacob's staff, and they report that, by actual measurement, we are now annexing the Gulf of Mexico at the rate of two hundred and sixty-two feet a year, and were only four thousand five hundred years in filling up the thirteen thousand five hundred square statute miles of the delta—an area greater than that of Belgium or Holland.

' Lyell's Niagara excavation is another piece of antiquated engineering of the same kind. With marvellous perseverance he has kept the present little river digging and boring a channel for itself back from Lewiston to the Falls, through the hard, solid rock, for thirty-five thousand years, at the rate of a foot ·a-year, when he (Lyell) knew well enough that before the rock hardened it was a mortar soft enough to mould around the most fragile shells, and that there was a big old Palæozoic ocean on the doorsteps, with the snow-shovel and broom with which he had cleared out a pathway for the Tennessee, and all the other rivers of that era, and, as everybody acknowledges, had actually cleared out the St. David's Channel. Was it to be supposed that, after such a length of excavating, he would stop at Lewiston, and leave the poor young river to shovel away the few remaining miles? Professor Christy shows he was no such churl. The process of excavating the Niagara channel did not occupy the old ocean as many hours as Lyell demands years for the job.

' It is now generally acknowledged that America does not stand on the same geological level with the Old World, so called—being a whole geological era older than it. Europe and Asia are now in their quaternary period, while we are yet in our tertiary. The plants and animals of the European tertiary are now alive in our woods and seas. The grand significance of this discovery is the demonstration of the falschood of the assumption—indispensable to geological chronology— that similar fossils prove similar strata, and that similar strata are

or twelve years subsequently, an old Gaulish history was re-edited, from which it appears, that during three years, long after Julius Cæsar, viz. in A.D. 458–460, the district was convulsed with violent and continued volcanic eruptions, and streams of lava carried destruction before them!* Too much is built upon negative evidence. It is from such ill-founded assumptions, and from partial or only average data, that all the

of the same age. Fossils similar to those of the European tertiary are now being found here in our thermal springs and on our coast, while geological cycles have passed since those of Europe were buried many a fathom deep.

'But the entire reversal of the whole fabric is completed by the dissolution of its foundation—the granite—which has just been discovered to be an aqueous formation from the overlying sedimentary rocks, instead of an igneous rock, from the abrasion and solution of which all the upper rocks were formed. It is impossible to imagine any discovery more fatal to the regular geological theory, or more mortifying to its expounders. Only last night, in a lecture here (Chicago) on 'Man's place in the Creation,' Agassiz began by stating that 'it is probable our globe was once in a state of igneous fusion,' &c., and tracing its progress towards a state habitable for man: in which lecture, however, he crushed to atoms the Development Theory, and demonstrated the unity of the Divine plan of creation from the earliest ages, making Atheism and Pantheism fall before crabs and worms. In fact, almost the only new truth demonstrated by geology is the recurrence of miracles, the repeated exercise of creative power on our globe. But those who grant this, as most first-class geologists do, cannot object to the probability of the miracles of Scripture. For if God was in the habit of working miracles here before man's advent, no reason can be shown why He should stop on our account, but rather additional reason why, by Divine works, He should demonstrate His presence to His children.

'We do not attach any great importance, however, to these geological proofs of natural religion, having very little confidence in so-called geological facts, most of which derive all their significance from the igneous theory, which is now being extinguished by the experiments of Daubrée and Rose.'

* *Quarterly Review*, October 1844.

attempts at chronological measurements are made in geology.

The date fixed by Sir Charles for the Clyde canoes, already noticed, receives rather a rude shock from the suspicious circumstance of a *cork*-plug being found, filling up a round hole in one of them; especially, as it appears from Professor W. King's statement (which I quote from the *Christian Observer* of May 1863), that this occurred, not in one of the more finished boats, but in one considered the most ancient, 'in a large rude one hollowed out of the trunk of an oak, which, when found, was quite black, as hard as marble, and very heavy.' He adds: 'Canoes formed of the solid stem of an oak were in use in Ireland about 200 years ago. It is difficult to arrive at any other conclusion than that the Glasgow canoes, instead of being several thousand years old, are comparatively modern, the one containing the cork-plug having been in use in the first or second century.'

As an illustration of the important influence which a better knowledge of the history of past events and customs might have upon some of our scientific conjectures, I refer my readers to an article in *Macmillan's Magazine* (September 1862), entitled *The Hand of Man in the Kirkdale Cavern*, by Mr. John Taylor. The practice of the Druid magicians of using the bones of many animals, and especially of the hyæna, for medicinal purposes, as related in Pliny's *Natural History*, is there applied to account for the collection of bones, particularly of the hyæna, in this cavern; the entrance to which is said to be too small for any living animal of the species to have dragged in its supposed victims. Also, the remains discovered in the greatest abundance are just those very parts of the animal which Pliny states were most used by the Druids. The idea is that the cave was the abode of these Druid magicians, and that the bones were imported from countries where the animals lived, for magical, medicinal, and other purposes. The article is very curious, and well worth perusal.

(3.) Sir Charles's arguments, based upon the rate of upheaval of continents, are no sounder than those depending on the assumed rate of deposits. He bases some of his calculations upon the assumption that two feet and a half in a century is the rate of upheaval, for immense periods of time, in regions of which he is writing. After his own admissions regarding the fluctuations of level in various parts of the world— for example, the Bay of Baiæ, Cashmere, Chili, and Peru (pp. 45, 46)—how can he possibly presume that there were no sudden upheavals, which by successive shocks might do the work in a few centuries, which, at his assumed uniform rate, would occupy as many thousands or tens of thousands of years?

One instance on this subject will suffice. At Cagliari, on the coast of Sardinia, have been found, about 300 feet above the sea, some sea-shells mingled with remains of pottery. It is supposed that the coast has been gradually upheaved, and that at the time when those shells belonged to living fish in the sea below, man must have been an inhabitant of the coast, and that those specimens of his pottery fell into the waters and mingled with the shells. But how are we to ascertain how long the coast has taken to rise these 300 feet? No help can be obtained from the phenomena of that country. So Sir Charles goes to Sweden, quite a different part of Europe, where the local change of level has been ascertained with some degree of exactness. The observations there extend, however, over only about one century, and are therefore no proper measure even in that country for thousands of years. Moreover, the movement in Sweden along the coast of the Baltic is found to be unequal, and also irregular at different times in the same place. The average rise is about $2\frac{1}{2}$ feet in a century. This then, strange to say, though only an average, and belonging also to a totally different country, and the result of only a single century of observation, he applies to Cagliari, and finds the age of the pottery to be 12,000 years! A process of reasoning this, so loose, so gratuitous, that one is really ashamed, for the cause of science, to say that a scientific man has produced it. No doubt he has prefaced his calculation with an 'if,' 'if the coast has risen $2\frac{1}{2}$ feet in a century.' But when his gratuitous hypothesis leads to a result so contradictory to the Bible, what ground had he to adopt it at all? Why did he not quote his own book, where in another place he shows, that in the west coast of South America permanent upheavals of the land of *several feet at a time* have been experienced in recent periods? Why did he not attribute the 300 feet of Cagliari, in part at least, to such shocks, which would have brought his period within the Bible chronology? These are specimens of the manner in which objections are set up against the historical truth of the Holy Scriptures. There is this benefit, indeed, arising from such speculations,—they show how really unassailable the Sacred Volume is, as no better arguments can be produced against it.

(4.) Nor does his theory regarding the length of time necessary for the disappearance of species appear to be more than a simple assumption. Why should a species be supposed always to die out *gradually*, when it disappears from creation? No valid reason whatever is assigned for this. Our readers will be surprised at hearing Sir Charles's reason. 'We know how tedious a task it is in our own times, even with the aid of fire-arms, to exterminate a noxious quadruped—a wolf, for example—in any region comprising within it an extensive forest or a mountain chain' (p. 374)—a strange comparison, indeed, between man's puny powers and the forces in the natural world! The opinion formed by Mr. Prestwich—an authority of which Sir Charles himself speaks so highly (p. 103) —from a careful examination of the phenomena in France and England, is very different. It is to be found in his Report to the Royal Society, already alluded to. His impression is in favour of the abrupt disappearance of species; and he conceives that the whole facts show, that the epoch of extinct animals should be brought down nearer to our own time, rather than that man's first appearance should be put further back.*

* 'Whilst abstaining from any general hypothesis in explanation of the phenomena, there is, however, one point to which I must refer before concluding, although I cannot at present venture beyond a few generalities respecting it. It may be supposed, that in assigning to man an appearance at such a period it would of necessity imply his existence during long ages beyond all exact calculations; for we have

The discussion regarding the age of these works of art affords a curious instance of inversion in argument. The diluvium of geologists has been hitherto looked upon as ancient, simply because no remains of man, but only of huge land-mammals, had been discovered; but now that traces of man are found, it is most illogically inferred that man must have existed for ages before been apt to place even the latest of our geological changes at a remote, and, to us, unknown distance. The reasons on which such a view has been held have been mainly, the great lapse of time considered requisite for the dying out of so many species of great mammals, the circumstance of the smaller valleys having been excavated since they lived, the presumed non-existence of Man himself, and the great extent of the later and more modern accumulations. But we have in this part of Europe no succession of strata to record a gradual dying out of the species, but much, on the contrary, which points to an abrupt end, and evidence only of relative and not actual time; while the recent valley-deposit, although often indicating considerable age, shows rates of growth, which though variable, appear, on the whole, to have been comparatively rapid. The evidence, in fact, as it at present stands, does not seem to me to necessitate the carrying back of man in past time, so much as the bringing forward of the extinct animals towards our own time; my own previous opinion, founded upon an independent study of the superficial drift on pleistocene deposits, having likewise been certainly in favour of the latter view. There are numerous phenomena which I can only consider evidence of a sudden change, and of a rapid and transitory action and modification of the surface, at a comparatively recent geological period—a period which, if the foregoing facts are truly interpreted, would seem, nevertheless, to have been marked, before its end, by the presence of man, on a land clothed with a vegetation apparently very similar to that now flourishing in like latitudes, and whose waters were inhabited by testacea of forms also now living; while on the surface of that land there lived mammalia, of which some species are yet the associates of man, although accompanied by others, many of them of gigantic size, and of forms now extinct.'—Joseph Prestwich, Esq. F.R.S. F.G.S., *On the Occurrence of Flint Implements associated with the remains of animals of extinct species in beds of a late geological period, in France, at Amiens and Abbeville, and in England at Hoxne.*—*Phil. Trans.* 1860, p. 277.

the historical period, because his remains are found in beds which have been pronounced to be pre-Adamite according to the received chronology.*

I cannot hesitate to give a verdict against the conclusion to which Sir Charles Lyell comes in his new work. The facts which have been here brought together from various sources are no doubt very interesting. But the hypothesis built upon them is crude and hastily constructed.† It is observable, that throughout the book there is no allusion whatever to the contradiction between the result he arrives at and the narrative of Holy Scripture; no reference to there

---

* See this pointed out in an excellent article on the subject in *Blackwood's Magazine*, October 1860.

† Mr. Pattison gives the following sketch, which in itself bears as much *à priori* probability on the face of it as Sir Charles Lyell's view; and it has the advantage of bringing all the events we have enumerated within the range of Scripture chronology:—

'There is nothing in the ascertained facts of geology, nothing in the exhaustive volume before us, to forbid the hypothesis, that at some period after the final retreat of the glaciers [which the glacial theory shows preceded the appearance of man], man found his way into these regions as a wandering hunter, probably from a distant geographical centre; that he resorted to these parts at intervals during several thousand years; that its pebble-beds afforded him implements, and its grassy beds abundant game; that in the intervals of his occupation the earth was rent, in connexion with volcanic action dying out in the Eifel and Auverge, floods occurred, the loose materials of the surface were washed into the crevices, or spread out in heaps; that many of the great mammals became extinct, some so lately as the mammoth, whose flesh was found in ice at the mouth of the Neva. For upwards of 4000 years all things were in course of becoming what they now are; and what they so became they have remained, save surface accumulations and minor changes, for the last 2000 years and upwards. For aught that geology or palæontology has yet to show, this is as valid an explanation of the phenomena as that which, under the semblance of indefiniteness, is carefully definite for a long time before Adam. If it is physically and philosophically possible to inter-being any difficulty, much less any attempt at clearing up the wide discrepancy. We cannot too strenuously adhere to the principle, that Scripture and Science should be kept well apart in our processes of reasoning; that is, that no fact from one should be taken to fill up a break in the line of argument which is suggested by the other. But to say that results should not be compared at certain salient points, where the lines of argument touch or cross each other, surely betrays an indifference, if not treachery, to the real interests of truth. Such a view can proceed only from that false notion, which some are known to hold, that the earlier

calate all the epochs of man, shown in the monuments of the globe itself, within the compass of the years assigned to the same occurrences by the received interpretation of Scripture, my task is done. I claim the verdict of " Not proven " on the issue raised.'

The following extracts from a more recent pamphlet by Mr. Pattison, *New Facts and Old Records* (1867), will be read with interest :—

' The comparative insignificant dimensions of the stratum containing human remains is shown by any geological table expressing the thickness of the several strata .... The thicknesses of British strata containing traces of former life are as follows :—

	Feet		Feet
Post-Tertiary:		New Red Sandstone	1,200
With human remains	50	Permian	1,000
Without „	50	Coal Measures	17,000
Tertiary	1,800	Devonian	4,000
Chalk	1,200	Silurian	12,500
Oolites	4,200	Cambrian	26,000

.... Thus, after geology has led us upwards through inconceivably long ages of God's working, over mountain-piles of strata, across intervals of long repose, through cycles of extinct life, we come at length up into the last display of its treasures, and encounter, as it were, the fresh air of the upper world in the presence of the human race .... So the Bible introduces Adam to us as the crowning work of the creation, the beginner of human history. After this period geology and Scripture run parallel for a short time. Geology tells us of the flint-implement makers, that they resorted to caves and dens of the earth, and to lake and river banks. Archaeology takes up the

chapters of Genesis are a collection of merely human legendary fragments, unhistorical, and therefore of no real value—an assumption utterly at variance, as I shall show in my Second Part, with the character which our Lord and His Apostle have stamped upon this portion of Holy Scripture, by the perpetual use they make of it to illustrate and establish sacred truths. The mind imbued with such loose and, in fact, untenable views regarding these chapters, feels itself under no restraint, and, deluded into a false liberty, is carried along through the regions of speculation without check. Another marked and dangerous effect of this state of mind is seen in the very slender premises from which men are sometimes contented to draw a great result. Were a due regard, indeed *any* regard, enter-

wondrous tale, and informs us that they had successors who used similar tools, but somewhat improved, polished, and varied by foreign material, until history dawns on the haunts and homes of our Celtic or pre-Celtic ancestors. Geology hitherto has only found traces of man in Western Europe, afar from the primeval seat of patriarchal revelation . . . . The circumstance that the earliest traces of mankind in Europe are traces of barbarian life, has been used by some modern speculators as evidence that such was the original condition of man. If we could be sure that we have exhumed the works of the first men, and had no other testimony about them, this would be admissible. But as we have no reason to believe these assumptions, but the contrary, we may reject the conclusion . . . . Scripture declares that men wandered away from the institutions and knowledge connected with the true religion, and became wild and Pagan . . . . The flint-implement makers, the lake-dwellers, the cave inhabitants of the West, were contemporaneous with partial civilization in the East . . . . Geology requires no pre-Adamite men, nor does archæology or ethnology demand that any other human race than the present should have lived on the earth . . . . All the occurrences posterior in date to the first appearance of the remains or works of man may find a place in the interval of 4000 years between the Creation and the opening of the history of Europe.'—(Pp. 10–19).

tained for the momentous importance of the result, when it appears to contravene the Holy Scriptures, the candid reasoner would in such a case abstain from drawing any inference at all where the data are uncertain and scanty. But the tendency to generalize is often too strong to endure such restraint. It seems to be considered necessary, that *some* result *must* be arrived at, and *must* be announced; and, though its announcement may be accompanied with the caution that the facts upon which it is based are few, and not altogether trustworthy, yet it goes forth as a *result*, the caution is too often forgotten, and thus uncertainty is more mixed up with the conclusions arrived at than most persons suspect.*

The work of Sir John Lubbock, *Pre-Historic Times*, is

* The following telling facts are extracted from the *Medical Times and Gazette*, New Series, part cxcix., Feb. 1867, p. 120:—

'*The Romans at London Wall.*—Our readers have probably seen from time to time notices in the daily papers of excavations that have been lately carried on in the vicinity of London Wall. A detailed and authentic account of the very remarkable human remains has been presented to the Anthropological Society of London by Lieut.-Colonel Lane Fox, and we hear that a description by Mr. C. Carter Blake of the remains of animals occurring in the excavation is to appear in the *Geological Magazine*. Through the kindness of the latter gentleman we have been allowed to inspect some portions of the treasure-trove, and the scientific interest which attaches to them will be a sufficient apology for their notice here. It appears, from Lieut.-Colonel Fox's account, that a section of the soil excavated shows at a depth of from 17 to 22 feet a bed of gravel similar to Thames ballast. On this rests a bed of peat from 7 to 9 feet thick, and above all are the modern remains of London earth, composed of the accumulated rubbish of the city. In excavating the peat it was found to be interspersed with numerous wooden piles, arranged in clusters, rows, and curves; these averaged from 4 inches to a foot square: the lower ends, which are pointed, were inserted one or two feet in the subjacent gravel, and the tops, which had rotted off, projected about 2 feet under the peat.

another like this of Sir Charles Lyell's, most interesting in the large amount of facts which it brings together. But the same vague inferences are drawn with a sweeping hand regarding the vast antiquity of the human race. And he conceives them confirmed by M. Bunsen's theory of languages, Egyptian monuments, the Negro race, the

These piles were roughly hewn, as if with an axe, and have been identified as of oak and elm. In the gravel no organic remains were found; but the peat was full of bones, and of works of Roman art — the former broken and black from having lain in the peat for centuries. Several human skulls were found, together with the remains of horse, ass, red deer, wild boar, wild goat (bouquetin), dog, roebuck, and several species of *Bos*,—viz. *Bos primigenius, B. longifrons, B. frontosus,* and *B. trochoceros.* The works of human art were various. Besides the piles, there were planks found, one plank having several Roman nails in it. There were also a quantity of red Samian pottery, some of it bearing men and animals cast in relief; bronze and copper pins, iron knives, tweezers, iron shears, a piece of polished metal mirror, so bright that the face may be seen in it, and a quantity of leathern soles of shoes or sandals, some much worn, some thickly studded with hobnails, the *caliga* of the Roman legion. Besides these, there were articles of a ruder construction, which may reasonably be supposed to have been of British manufacture — spear-heads (?) made of the metacarpals of the red deer and *Bos longifrons*, and skates formed of the metacarpals of a pony or donkey. One of the most remarkable points connected with this excavation is the great thickness of the peat, the evidence given by a section of its upward growth, and the fact that from its lowest to its highest level it contains Roman remains. Thus, at a foot and a half above the gravel was found a kitchen midden of oyster and mussel-shells, with Roman pottery and a Roman *caliga*. Similar refuse-heaps occurred at higher levels in the peat, and at its very top Roman remains of the same kind were found. It is evident, therefore, that the enormous growth of from 7 to 9 feet of peat must have occurred during the 400 years of Roman occupation. Sir Charles Lyell, following M. Boucher de Perthes, would have estimated the probable growth of peat during that period at about four inches. To quote Colonel Lane Fox, the history of the growth may be read by the sections, as follows:—

1. Oak piles driven into the gravel, the tops of which rotted off at the surface before the peat had grown more than two or three feet.

trees of Denmark, the deposits of the river Tinière, the Nile-deposits of Mr. Horner, the Mississippi delta, the alluvium of the Somme,—all of which are shown in this treatise to be either untrue or insufficient to support the conclusion drawn, in places which may at once be found by referring to the Index. He concludes

2. A kitchen midden deposited on peat a foot and a half thick during the Roman period. This may, or may not, have been contemporaneous with the first piles. 3. A growth of peat of one or two feet above this kitchen midden, and other piles then driven in. 4. A kitchen midden with *Bos longifrons* and Roman pottery at three and a half feet. 5. Another growth of peat and another kitchen midden at six feet; and, lastly, Roman remains at the very top.

With regard to the use of the piles found driven in the peat, there can be no doubt, from the kitchen middens in their vicinity, that they must have been the foundations of human habitations. The remains of no superincumbent architecture were found above them; and whether they were the work of Roman or British workmen remains a question. Similar piles and remains have been observed in excavations in other parts of the City, as near the Mansion House, in the line of the old Wall Brook and north of the Bank. Colonel Lane Fox suggests that they may be the remains of the British capital Cassibelaunus, which Cæsar says he found situated amongst woods and marshes. Another point of great interest is, that there can be now no doubt that the giant bull, *Bos primigenius*, the congener of *Elephas primigenius*, *Ursus spelæus*, and *Cervus megaceros*, was not extinct at the time of the Roman occupation of England. Reasons for accepting this conclusion were stated in this journal some years ago (June 28, 1862), and it must be hereafter an accepted fact. The tide of discovery seems, on the whole, to be setting in favour of a shortening of the geological periods, which the observations made a few years ago seemed to render necessary, in the interpretation of human remains then first brought to light. If it be so, modern scientific teaching may yet be brought into accordance with the history and traditions of civilized man.'

As a specimen of the hazard scientific men sometimes run of being grossly imposed upon, I give the following extract from the report of a scientific meeting held at Leeds since the last edition of this treatise was published:—

'Professor Owen said, that some time ago he was sent for to the north, to examine a fossilized tree which had been found in digging

his chapter on this subject by an account of the interesting discussion, which has been going on of late years, regarding the date of the Glacial Period— *assuming* that the human race existed then. The rein which he gives to his imagination may be seen in this short sentence,—'It may be doubted whether even geologists yet realise the great antiquity of our race.'*

<small>Species of the Six-days distinct from pre-Adamite species.</small> 8. A difficulty has been started against the literal interpretation of the first chapter of Genesis, to which I have already adverted, but on which I will now enlarge, because I think it furnishes an illustration of my main argument, by showing, to say the least, that the positive statements which are sometimes put forward in geological

the Jarrow Dock, which bore undoubted evidence of having been cut by human hands. It was supposed to be a most important discovery, as showing the antiquity of the human race; and at first everything appeared satisfactory. On prosecuting his inquiries, however, he learnt that one of the navvies not then on the works was said to have discovered a similar tree in another part of the dock, which he cut to lay down a sleeper. The man was sent for, and at his arrival he declared that the tree pointed out was the one he had cut. It was endeavoured to be explained that that was impossible, as the place had not been excavated before; but, looking with supreme contempt upon the assembly of geologists and engineers, the man persisted in the identification of his own work, and exclaimed, 'The top of the tree must be somewhere;' upon which he (Professor Owen) offered half-a-crown to the first 'navvy' who would produce it. Away ran half-a-dozen of them, and in a few minutes they returned with the top. This explained the mystery. The man had cut off the top with his spade to make way for a sleeper; the stump afterwards got covered up with silt, and on being again uncovered it was supposed to be a great discovery. Never had he so narrow an escape from introducing a "new discovery" into Science, and never had he a more fortunate escape.

* *Pre-Historic Times.* Second Edition, p. 410.

theories are often very untrustworthy, and cannot be received as evidence against the Scriptures.

It has hitherto been an established principle with physiologists, that individuals propagate in the same species; and never out of one species into another. Mr. Darwin's new theory upon the subject cannot yet be said to have displaced the old one. Sameness of species is, therefore, considered necessary to physical continuity; and physical continuity, on the other hand, has generally been inferred where sameness of species has been found at two different epochs. About forty years ago it was maintained by Sir Charles Lyell that the researches of M. Deshayes, a conchologist, of Paris, had proved that multitudes of pre-Adamite fossils in the upper (or Tertiary) strata, are precisely the same as species now in existence.* Shells, found fossil in the rocks, were pronounced by the scientific conchologist in many instances to be precisely the same as species now living in the neighbouring seas. Here, then, we seemed to have a chain of living links between the present times and the pre-Adamite periods. It would follow, that creatures must have been in existence, on some parts of the earth, during the six days' work, which had lived before that time and partook not of the creation of that period. Dr. Pye Smyth attempted to solve this difficulty, which seemed utterly opposed to the letter of Scripture. His solution was simply, that the word translated *earth*, in the Old Testament, is as

---

* Indeed, upon this hypothesis the designations of the several divisions of the Tertiary Period were framed—the *Eocene*, or dawn of recent (species); the *Meiocene*, or minority of recent (species); and the *Pleiocene*, or the majority of recent (species).

often rendered *land;* and that, therefore, it might be so rendered in the account of the creation of plants and animals during the six days. The Mosaic account would then refer only to a local creation of certain races of animals and plants at the epoch when man was formed, and in the country which surrounded Paradise, and not to a creation taking place over the whole earth. This scheme of limiting the creation of the six days to the precincts of Paradise satisfied many, under the circumstances.

But this identity of species was afterwards called in question. Further investigation and a large multiplication of facts seemed to show that this conclusion was premature. The late M. d'Orbigny, another eminent conchologist, re-examined the fossils and many others, and came to the conclusion that between the termination of the Tertiary Period and the commencement of the Human or Recent Period, there is a complete break. Although five in every seven *genera* are the same in the recent as in the previous period, he asserted that there is not a single *species* common to the two periods.[*] The difficulty therefore vanished. No resort to the forced expedient of Dr. Pye Smith seemed called for. What was at one time a formidable objection, apparently, became one of the most striking confirmations we could possibly have, that Scripture and Science are never at variance when interpreted aright. But Sir Charles Lyell, in his recent work on *The Antiquity of Man*, has re-asserted his original views (see note on

[*] See an account of M. d'Orbigny's discoveries, in the articles of the 'Pre-Adamite Earth,' Vol. xii. of Lardner's *Museum of Science and Art*.

SPECIES OF THE 'SIX-DAYS' AND BEFORE ADAM.

page 83 of this treatise), notwithstanding the opposition it has met with from M. d'Orbigny and other eminent conchologists. These are questions which can be decided only by scientific observers who have devoted their lives to these peculiar studies. We must not be surprised at these reverses. Science in her onward progress has often had to encounter them. It has often happened, too, that facts discovered and embodied in erroneous theory, and even errors made, have had their importance in leading to truth at last. But this is clear, that nothing has been proved in this case against Scripture, and we should learn never to allow a theory, however plausible, to stand forth as an argument against the infallibility of the Word of God. If eminent men differ so materially on the results they come to, the conclusion is clear that those results should never be produced in the controversy set up between Scripture and Science. It is not at all impossible that even Sir Charles Lyell may change his opinion on this subject, as he has already done in more cases than one. He followed Hutton most devotedly in his reasoning on the igneous origin of granite; but has of late come round to the new and opposite view, or rather the old neptunian view revived in opposition to the plutonic, that granite is of aqueous origin. The remarkable results of the late deep-sea dredgings in the North Atlantic to nearly three miles, will most likely make a great change in his views on the subject of climate as determined by the distribution of fossils. For Dr. Carpenter informs us, that at very great depths the sea is lower in temperature than even 30° Fahr., far into the temperate zone; that is, as far as the dredging had been then carried; that under-currents of cold, and therefore heavy, water, flow down from the polar regions; and that it is most likely that an actual continuity exists between the arctic and antarctic fauna, even under the warm surface of the tropical ocean! He says, that these results materially affect the conclusion to which geologists have hitherto come, regarding the general prevalence formerly of a glacial climate. And he adds, that 'it may be considered as having been now placed beyond a reasonable doubt, that a glacial submarine climate may prevail over any area, without having any relation whatever to the terrestrial climate of that area.'* Moreover, on the subject of development, which I take up in my next example, Sir Charles Lyell has entirely gone round from one opinion to the opposite. In the earlier editions of his *Principles of Geology* he strenuously opposed Lamark's theory, showing that no doubt the facts brought forward illustrate 'the unity of plan that runs through the organization of the whole series of vertebrated animals;' but that 'they lend *no support whatever* to the notion of a gradual transmutation of one species to another.'† This is in direct opposition to the theory of Mr. Darwin, which Sir Charles Lyell *now adopts*, as may be seen in his work on the *Antiquity of Man*. No doubt we are all liable to make mistakes, and to assert at one time that '*no support whatever*' is to be derived from certain phenomena for a theory under

---

* *Proceedings of the Royal Society*, No. 121, November 18, 1869. 'Preliminary Report of the Scientific Exploration of the Deep Sea,' pp. 475, 476.

† *Principles of Geology*. Third Edition, p. 102.

consideration, which those phenomena are, at a subsequent period, pronounced to establish. But such a revolution in our opinions should, to say the least, make us diffident in our assertions and deductions from natural appearances for the future; especially when they go counter to clear statements in Holy Scripture.

9. Mr. Darwin's work, on the *Origin of Species*, suggests another illustration of my argument. His view appears to be this—If all creatures, plants and animals, were to bear fruit, since the increase whenever it does take place is by multiplication, and not by the simple substitution of one individual for another, the vegetable and animal kingdoms would be soon so overstocked that there would not be standing room on the earth's surface. There is, therefore, constantly going on in the world a Struggle for Existence; and those individuals survive and propagate which have certain natural advantages which enable them to maintain their position. This process he calls Natural Selection. When an individual pair produces other individuals, these, as is well known, often exhibit varieties even within the limits of the same species. Naturalists have hitherto considered that these varieties in the long-run disappear; and that species run on in lines absolutely invariable in their main features. Just as with the planets in their orbits, which are perpetually deviating from their average ellipse: the variations neutralise each other in a vast number of revolutions, the average ellipse is absolutely invariable, and the configuration of the system remains unaltered even after the lapse of many ages. Mr. Darwin, however,

Origin of Species.

considers that this is not the case with species, but that the varieties accumulate their effects in the course of immense periods, and give rise to other species. Of the individuals which are the offspring of a single pair, he conceives that those are most likely to survive which differ most from each other, as they will interfere less with one another in the struggle for existence than with the intermediate ones, which are nearer to them in character. Hence Natural Selection seems likely to promote gradual divergence. He gives us no idea whatever how it has come to pass that this process should happen to produce both a male *and* a female : without which the new species would at once die out; as it is well known that hybrids, the offspring of parents belonging to different species, have no offspring themselves. Even if we could conceive that the individual specimen of a new species, supposed to be produced by natural selection, so closely resembled the species from which it came, that it might have fruitful offspring from another individual of that species, the offspring would be different to both parents, as they are of different species; and unless the individual members of the offspring of this abnormal pair intermixed, without mixture with members of the original species, to which they are so closely allied—a thing most improbable,— no permanent species, such as we see now, could have been produced by this process.* This hap-hazard mode

---

* As near an approach to this as can well be conceived is related by Colonel Humphreys in the *Philosophical Transactions*, 1813, Part I. (quoted by Professor Huxley in his *Lay Sermons, Addresses, and Reviews*, pp. 290-295). An extraordinary birth took place among sheep, which was taken advantage of, by artificial breeding, and a new

of peopling the world with creatures, appears to my mind to lead to confusion, and not to the beautiful order we see actually prevailing.

The term 'Natural Selection,' which some have admired for its power of expressing the meaning to be conveyed, seems to me objectionable. It appears to be chosen to exclude all idea of creation—the creation of species. The act of selection implies an agent. But who is the agent here? Were the term human-selection or divine-selection, we could easily give a reply. The term Mr. Darwin has chosen implies that Nature is the Selector. But who is Nature? He might much better have called it at once Chance-Selection; because the term would have been just as good, and, indeed, much better, because it would bring out from its concealment the palpable contradiction involved in the term he adopts. Perhaps he would reply, 'No, there is no chance in this: in the struggle for existence those who survive survive by the laws of nature, and may be said to be selected by those laws.' But this is arguing in a circle. Laws! he has not shown what the laws are; they are but laws in the sense of an order of classification. His whole reasoning is only a piece of taxonomy, and nothing else. There is no tracing whatever of the *origin* of species. He gives in no single instance any physiological cause for a new species springing up in this way; he only dignifies the

race of sheep was raised, called the Ancons, which retained its characteristics; till, however, it was neglected, when, by breeding again with ordinary sheep, the peculiarities disappeared. The power of breeding, moreover, with other sheep showed that, notwithstanding the peculiarities of the new breed, it was not a new species.

animals and plants which survive in the struggle for existence as the offspring of the personified abstraction, Natural Selection. His theory only shows how they survive destruction; but in no respect whatever explains how, differing as they do from their parents, they came into existence endowed with peculiarities. His language really amounts to no more than saying, What is, is.

Mr. Wallace, the contemporary propounder of this theory with Mr. Darwin, and a strong advocate for it, considers that Natural Selection is insufficient to account for the development of man; and that certain phenomena suggest that 'a superior intelligence has guided that development in a definite direction, and for a special purpose.' These phenomena are as follows:—

'The brain of the lowest savages, and, as far as we yet know, of the pre-historic races, is little inferior in size to that of the highest types of man, and immensely superior to that of the highest animals; while it is universally admitted that quantity of brain is one of the most important elements which determine mental power. Yet the mental requirements of savages, and the faculties actually exercised by them, are very little above those of animals.'—'They possess a mental organ beyond their needs. Natural Selection could only have endowed savage man with a brain a little superior to that of an ape; whereas, he actually possesses one very little inferior to that of a philosopher.' Again, 'the soft, naked, sensitive skin of man, entirely free from that hairy covering, which is so universal among other mammalia, cannot be explained on the theory of Natural Selection. The habits of savages show that they feel the want of this covering, which is most completely absent in man exactly where it is thickest in other animals. We have no reason whatever to believe that it would have been hurtful, or even useless to primitive man ; and under these circumstances, its complete abolition, shown by its never reverting in mixed breeds, is a demonstration of the agency of some other power than a law of the survival of the fittest in the development of man from the lower animals.' He draws similar arguments from the structure of the hands and feet ; also the human larynx, 'giving the power of speech, and of producing musical sounds, and especially its extreme development in the female sex, as shown to be beyond the needs of savages, and from their known habits impossible to have been acquired, either by sexual selection or by survival of the fittest.'* These are most important admissions, and from such a quarter.

Mr. Darwin's speculations appear to be divisible into two parts : first, as to species being mutable or not ; and secondly, if they be so, whether all creatures have descended, by natural selection, from a primordial germ. It is for naturalists to examine his arguments and to decide upon the theory of the mutability of species. This may be true or not. It is not with this result, true or false, that I have now to do,† but with

---

* See the whole of the chapter on ' The limits of Natural Selection as applied to Man ' in Mr. Wallace's book, *Contributions to the Theory*.

† ' If it does not diminish, but only augments the wonder of Organic Life, that it has been so contrived as to be capable of propagating itself, neither would it diminish that wonder, but rather enhance

his speculations arising from this new principle, and which appear to me to be directly contradictory to Scripture. These speculations are partly involved in the very title of his book—*On the* ORIGIN *of Species*. But they are fully expressed in the following words :— 'I believe that animals have descended from at most only four or five progenitors, and plants from an equal or lesser number. ... I should infer from analogy, that probably all the organic beings [plants and animals] which have ever lived on this earth have descended from some one primordial form, into which life was first breathed' (p. 484) Even God's rational creature is included in this category :—' Light,' he says, ' will be thrown on the origin of man' (p. 488). Man

it to an infinite degree, that Organisms should be gifted with the still more wonderful power of developing Forms of Life other and higher than their own. So far, therefore, as belief in a Personal Creator is concerned, the difficulties in the way of accepting this hypothesis [of the origin of species] are not theological. The difficulties are scientific. The first fundamental difficulty is simply this,—that all the theories of development ascribe to known causes unknown effects —unknown as regards the times in which we now live, and unknown so far as has hitherto been ascertained in all the past times of which there has been any record. It is true that this record—the geological record—is imperfect. But, as Sir Roderick Murchison has long ago proved, there are parts of that record which are singularly complete, and in those parts we have the proofs of Creation without any indications of development. The Silurian rocks, as regards Oceanic Life, are perfect, and abundant in the forms they have preserved, yet there are no Fish. The Devonian Age followed, tranquilly, and without a break; and in the Devonian Sea, suddenly, Fish appear—appear in shoals, and in forms of the highest and most perfect type. There is no trace of links or transitional forms between the great class of Mollusca and the great class of Fishes. There is no reason whatever to suppose that such forms, if they had existed, can have been destroyed in deposits which have preserved in wonderful perfection the minutest organisms.'—Duke of Argyll, *Primeval Man*, pp. 43-46.

comes, too, in the following passage, into this chain of being:—'The framework of bones being the same in the hand of man, wing of a bat, fin of a porpoise, and leg of a horse—the same number of vertebræ forming the neck of the giraffe and of the elephant—and innumerable other such facts, at once explain themselves on the theory of descent with slow and slight successive modifications' (p. 479).

But, 'What saith the Scripture?' The origin of *some* of the plants and animals which have lived on the earth, and that of man, is thus described:—

Gen. i. 11. 'And God said, Let the earth bring forth grass, the herb yielding seed, *and* the fruit tree yielding fruit after his kind, whose seed is in itself, upon the earth : and it was so. 12. And the earth brought forth grass, *and* herb yielding seed after his kind, and the tree yielding fruit, whose seed *was* in itself, after his kind : and God saw that it was good . . . . 20. And God said, Let the waters bring forth abundantly the moving creature that hath life; and let fowl fly above the earth in the open firmament of heaven. 21. And God created great whales, and every living creature that moveth, which the waters brought forth abundantly, after their kind, and every winged fowl after his kind : and God saw that it was good . . . . 24. And God said, Let the earth bring forth the living creature after his kind, cattle, and creeping thing, and beast of the earth after his kind : and it was so. 25. And God made the beast of the earth after his kind, and cattle after their kind, and everything that creepeth on the earth after his kind; and God saw that it was good. 26. And God said, Let us make man in our image, after our likeness . . . . 27. So God created man in his own image, in the image of God created he him; male and female created he them.'

To Mr. Darwin's 'mind it accords better,' he says, 'with what we know of the laws impressed on matter by the Creator, that the production and extinction of the past and present inhabitants of the world should have been due to secondary causes, like those determining the birth and death of the individual. When I view all things,' he adds, 'not as special creations, but as the lineal descendants of some few beings which lived long before the first bed of the Silurian system was deposited, they seem to me to become ennobled' (pp. 488–9). But if God has spoken, is not this language in the highest degree presumptuous?

Scripture and Science, then, as the latter is made to speak in these speculations, are decidedly at variance. But the fallacy of the naturalist's reasoning, I think, can easily be pointed out. Let us grant, for the sake of argument, that the great principle advocated in Mr. Darwin's work is true, and that those barriers do not exist which naturalists have hitherto supposed to be set up to prevent species losing their specific characteristics. The speculations which he adds, regarding the descent of all organized beings from a single primordial form, are altogether gratuitous, and form no part whatever of the theory itself. What brought the primordial form into existence? The pushing back of the first appearance of this form further and further into past time, ages before ages and ages before them, does not get rid of the question, How came this form into existence? a form, too, possessing such marvellous properties, as to give birth to all the varieties of organization which the vegetable and animal kingdoms exhibit. God must have created it. If, then, the Almighty created one such form, why could He not have created several? What necessity is there in the nature of things for tracing up the genealogy of all organic beings to *one* form only? All that the theory, if demonstrated, can establish is, that when God brought organized creatures into existence, He did not endow them, as naturalists have hitherto supposed, with that principle of permanence which has been designated the Immutability of Species.*

Another answer is this. Mr. Darwin appears to consider analogies in organic structure to be indications of identity of parentage, if we go back a sufficient number of generations. But this idea precludes the possible and highly probable one, noticed that the Almighty has sometimes seen fit to work according to a plan. The analogies he traces are, moreover, sometimes very distant from the end in view.† I have already noticed the instance he brings forward of the same number of cervical vertebræ being in the giraffe and the elephant. But if he considers this analogy important for his theory of descent, where does he

---

* That Mr. Darwin's conclusions may be sometimes hastily drawn, is very evident from the favourable light in which he views Mr. Horner's deductions from Nile-mud deposits, regarding the age of the human race (p. 18).

† For example, see the following:—' The illustration of the swim-bladder in fishes is a good one, because it shows us clearly the highly important fact, that an organ originally constructed for one purpose, namely flotation, may be converted into one for a wholly different purpose, namely respiration. The swim-bladder has also been worked in as an accessory to the auditory organs of certain fish, or, for I do not know which view is now generally held, a part of the auditory apparatus has been worked in as a complement to the swim-bladder. All physiologists admit that the swim-bladder is homologous, or "ideally similar," in position and structure, with the lungs of the higher vertebrate animals : hence there seems to me to be no difficulty in believing that natural selection has actually converted a swim-bladder into a lung, or organ used exclusively for respiration.' —Pp. 190, 191.

get the more numerous vertebræ in the neck of the plesiosaurus?*

Whatever be the fate of this new theory of the Mutability of Species, I do not think that it can in any way affect Scripture. It by no means leads of necessity to the speculations with which Mr. Darwin has fettered it; and, therefore, another instance is furnished in which an apparent discrepancy between Scripture and Science is swept away by pointing out the false conclusions which scientific men at times come to. It is observable from what opposite points the truthfulness of Scripture is attacked, and yet found invulnerable. Here Mr. Darwin contends from the physiological side for the unity of origin of all creatures, and therefore of all nations of the human family; whereas, strange to say, the authors of the *Types of Mankind*, to whom I have already adverted, and others like them, argue from the ethnological side for the very reverse of this, viz. the separate origin of each nation.

There is no doubt that the investigation of these

* 'The most anomalous of all the characters of Plesiosaurus Dolichodeinus is the extraordinary extension of the neck, to a length almost equalling that of the body and tail together, and surpassing in the number of its vertebræ (about thirty-three) that of the most long-necked bird, the swan; it thus deviates in the greatest degree from the almost universal law, which limits the cervical vertebræ of quadrupels to a very small number. Even in the camelopard, the camel, the lama, their number is uniformly seven. In the short neck of the Cetacea the type of this number is maintained. In birds it varies from nine to twenty-three; and in living reptiles from three to eight. We shall presently find in the habits of the Plesiosaurus a probable cause for this extraordinary deviation from the normal character of the lizards.'—Buckland's *Bridgewater Treatise*. First Edition, p. 205.

matters has added, and is still adding, vastly to our knowledge of the facts of nature. The authors of these works are, in this respect, doing service to the cause of truth. But it is not merely the accumulation of facts which we require, but also a sound induction from them. These two things are not generally the work of the same mind. Too hastily is a dagon-theory set up by the sanguine investigator, parts of which may be of excellent workmanship, and much of the materials be of the best description: but it is wanting in that principle of cohesion which can alone keep it standing under the shocks which stern truth gives it. Nothing has shaken my own confidence in the theories of our modern leading investigators so much as the reckless spirit of generalization which distinguishes this theory on which I am now writing. The spirit of sound philosophy appears to be forsaking some of our leading men. Dazzled by the multitude of facts which their investigations have brought and are continually bringing to light, they seem to me to lose that sober stability of reason which true philosophy needs. The spirit of speculation having too soon possessed them, they leap over many a chasm which would bring to a standstill men of sound reasoning powers. Enchanted by the simplicity of the idea their minds have conceived, they seem to forget the multitude of little guesses and assumptions which have paved their way over precipices which would have been fatal to the advance of sober-minded men. They are lost in a vast cloud of connected and unconnected facts, through which it is impossible for them to see; and, to use an expressive figure of the late Dr. Whewell, ' they suppose they are upon a mountain because they are in a mist.' The retrograde method of philosophizing which these speculators are in some cases manifesting is most un.-promising for the cause of sound Science. The refutation, for instance, which Sir Charles Lyell gave in the first edition of his *Principles of Geology* of Lamark's hypothesis of development, appeared to me at the time unanswerable, and does so still. But he has abandoned that ground now! and, in fact, under this new phase of the hypothesis, has adopted what he then stigmatized and also nicknamed as the 'energizing' theory. It presents, indeed, a perfect phenomenon of transformation to put side by side passages from the earlier editions of his *Principles of Geology* and his more recent work on the *Antiquity of Man*, which directly contradict each other. The natural result is inevitable; that no careful student, however much he may value the accumulation of facts which Sir Charles Lyell has ever been remarkable for bringing together, can put implicit confidence in him as a guide to sound conclusions. Men who invent or adopt theories full of loose joints, or joints merely in contact, but really unconnected, may happen to be very useful investigators of the facts of nature; but they are not philosophers. They forsake or overlook the great pattern of sound philosophizing, who, when the first thought of gravitation as the cause of the lunar and planetary movements possessed his mind, rejected it at once, because he found that the theoretical and calculated results in a particular case did not exactly agree; and, magnificent as the consequences of the theory have since proved to be—of which his penetrating genius must have had far more than a prescient glimpse—it was not till twelve years later, when improved data were furnished to him, that he took up the matter again, re-examined the effects of his first hypothesis, and found it to be true. To descend from such a type of trustworthy philosophizing to the speculators of the present day is most distressing; and, in some respects, painfully ludicrous. 'We require,' says an acute writer, 'a good amount of some sort of philosophy not to laugh when told that a duck, for instance, was not expressly intended to be a duck with a web-foot, that it might pleasantly move on the water, but that its forefathers and mothers a long way back began under pressing circumstances to get a duckish disposition, and by dint of endeavour for ages to try their chance of paddling themselves about on the pools of a puddly world, their efforts were at length quite rewarded, and resulted in a complete success—so remarkable, indeed, at last, that a generation sprang from them thoroughly equipped for the waters with web-feet, oily backs, boat-shaped bodies, spoony bills, and bowels to correspond with mudworms and duckweed.'* These wild hypotheses appear to me to be a disgrace to Science, and to indicate a downward movement, viz. from creation to development and evolution, which far outmeasures in extent the prodigious step upwards which Science made in the seventeenth century in the opposite direction, from the hypothesis of vortices to the theory of gravitation, as soon as the *Principia* appeared and made men ashamed

* *The First Man and his Place in Creation*, by George Moore, Esq. M. D., p. 36.

of having ever believed in such nonsense as the Cartesian system invented and taught: a system so absurd, that Whiston compared an adherence to it after Newton's was promulgated (and some did adhere to it), to eating old acorns rather than wheat flour.

10. The Origin of Man, God's intelligent creature on earth, both as to the mode of his first introduction into the world and his mental condition when introduced, is another question on which views have been broached contrary to Scripture, but which cannot stand the scrutiny of an impartial examination. Man's introduction into the world after the creation of other creatures is described in these sublime words in Genesis:—' And God said, Let us make man in our image after our likeness : and let them have dominion over the fish of the sea, and over the fowl of the air, and over the cattle, and over all the earth, and over every creeping thing that creepeth upon the earth. So God created man in his own image, in the image of God created he him ; male and female created he them' (Gen. i. 26, 27). 'And the Lord God formed man of the dust of the ground, and breathed into his nostrils the breath of life, and man became a living soul' (ii. 7). In the narrative of the six days, the inspired penman exhibits the Almighty in His true character : 'He spake and it was done, He commanded and it stood fast' (Ps. xxxiii. 9). But when He comes to the creation of man, the last and greatest of His works, the style is changed. The creation of man is described as if it had been the result of a special council, a council of the Sacred Trinity in Unity, as if peculiar importance attached to this work. 'And God said, Let us make man.' Under the Great

Origin of Man.

Supreme man was to be lord of the lower world. On him would depend its future well-being. Man was to be a new and distinguished link in the chain of being, uniting the irrational with the spiritual world, the frailty of the dust with the breath of the Almighty. How men, in the presence of such a description, do not quail before they venture to advocate such theories on man's origin as disgrace modern science it is difficult to understand. According to these theories Man has descended by natural generation, with all other organic creatures, vegetable and animal, from some one primordial form into which life was breathed. In the Struggle for Existence through ages countless beyond number, and according to the laws of Natural Selection, beings appeared at last in one chain of succession, of links all but infinite in number, which are described by one of the advocates of this monstrous hypothesis as 'the first men, or the first beings worthy to be so called.'* And in what character did these age-developed representatives, and, indeed, progenitors of the race now called Mankind, appear? Let the same theorist speak: 'The primitive condition of mankind was one of utter barbarism.' And he considers Adam to have been 'a typical savage.' And he adds, 'These views follow, I think, from strictly scientific considerations.'† In the last edition of his *Pre-Historic Times* he gives this view of the human race:—'The lowest races of existing savages must, always assuming the common origin of the human race, be at least as far advanced as were our ancestors

---

* *The Primitive Condition of Man*, by Sir John Lubbock, 1870, p. 325.

† Ibid. pp. 323, 361.

when they spread over the earth's surface.'* And he then goes through various races of savages, and points out something in which each was deficient; and comes to this deplorable conclusion, that primitive man was a creature who combined in his character all these negations! If this be Science, then indeed Scripture and Science are at variance.

But let us examine the facts. This these naturalists seem to have done to a certain point; but there to have stopped. The organic structure of man's body appears to resemble all other structures found throughout the animal kingdom. It is said, that organs found even in an undeveloped state in other animals, as not serviceable to them, are found perfected for use in the frame of man, and *vice versa*. This is the ground upon which this gratuitous hypothesis is built, of the community of the origin of man and the lower creation.†

* *Pre-Historic Times*, 2nd edit. p. 573.

† An account of various anatomical differences between man and the brutes will be found in *The First Man and his Place in Creation*, by G. Moore, Esq., M.D., Chap. II —a valuable and most interesting work. The following extract from the seventh chapter shows what troublesome differences exist, between man and even only the last link but one before him, for these speculatists to account for:—

'If it [the ape] had a more perfect hand, more approximating to our "instrument of reason;" if it had a less bearlike skull; if its teeth became more like aids to utterance and beauty, instead of being, as they are, mere instruments, like those of the dog, for cutting, tearing, and grinding hard food; if its spinal column were less curved; if its head were balanced on that column, as in man, and not hung on with the aid of special ligaments; if its hind paw were a real proper foot, fit to stand erect on, like man's, and not a compound convenience, neither quite hand nor foot; if, in short, its whole anatomy of bone and muscle were more like man's, the ape would be spoiled; and yet he would be not a whit more human in action or expression unless it had also a human mind.'

But cannot it have pleased Almighty God to have worked upon a plan? Why, if there be this similarity, must we necessarily assume that there is a relationship by generation among all the living creatures which have appeared upon the earth? Moreover, though there may be the resemblance spoken of, yet all the men, living or fossil, whose organization has been examined, are unmistakeably men and nothing else. No shades of difference gradually passing off from men to other creatures, the nearest resemblance to men, are to be found, living or in the earth. There is an abrupt break. The form of the skull, the oldest of all skulls which have been discovered in the earth's strata, and which was supposed by some to belong to an extremely ancient celt-using race, more allied to apes than men, is found to approach nearly to the Caucasian, the highest type of development*—a most damaging fact for this theory. Again, the difference in brain-power of a savage and a cultivated European is only 82 cubic inches compared with 94, while that of a gorilla is only 30 inches. How does this tally with theories of development, which make the savage a being of a far inferior order to the Caucasian, and which in no way can account for the storing up of power like this for future use? The savage is inferior simply because he does not use his brain to the extent of his capacity; not because he has not got it. His brain was given for higher purposes than he uses it for, and affords a proof that his first progenitors were of a higher stamp. Thus, even within their own chosen arena of debate—viz. the physical organization

* Lyell's *Antiquity of Man*, p. 89.

of man—their theory breaks down. They who would reason like philosophers on this subject, apart from the teaching of Scripture, will begin by taking man as he is; and will proceed by reasoning from this sure ground by legitimate steps, traced out by observation and experiment, to further views, if they are attainable. It has been well remarked, that 'it is only a low materializing philosophy which can regard the human body as man. By thinking and by loving man is man; and all the matter in the universe cannot produce a single thought, or perform a single act of love. It is the soul and spirit which make man what he is. If by digging in the earth we could find a fossil vestige of man's soul and spirit, then we might find man, as man is described in Scripture, but not till then,'*— and, 1 may add, as man is now. Can an example be produced, from any source, of a being in every respect like man, his higher nature alone being wanting? *When* did the immortal spirit which we now find in man first enter into the race? *How* did it enter? These surely are questions which we have a right to ask of those who so gratuitously set aside the distinct declaration of Revelation regarding the origin of man. The immeasurable difference between the lowest type of man and the highest of apes is, strange to say, acknowledged by one of the masters of this false philosophy.† And yet he argues thus: 'The structural

* Bishop Wordsworth's *Commentary on Genesis.*
† Professor Huxley. His words are these: He speaks of 'the great gulf which intervenes between the lowest man and the highest ape in intellectual power.'—*Man's Place in Nature.* Third edition, p. 102. Even regarding their physical structure, he says that 'the structural differences between man and even the highest apes . . . .

differences between man and even the highest apes
. . . are great and significant . . . It would be no
less wrong than absurd to deny the existence of this
chasm : . . . but do not forget that there is a no less
sharp line of demarcation, a no less complete absence
of any transitional form, between the gorilla and the
orang, or the orang and the gibbon.'* And on this
ground does he accept that alternative which is the
most contrary to reason, the most degrading to man,
and most offensive as regards Divine revelation, viz.
that all have had a common origin ! whereas the
legitimate inference would be, that, close as the resemblance of man's organic structure may be to that of
the highest brutes, their distance in more important
respects is so 'immeasurable,' that they can have had
no common origin ; and that, *consequently*, the various
species of brutes themselves, which are, if anything,
wider apart among one another than man in his physical structure is from the highest of them, must also
have had their *separate* origins ; that the declaration
of Holy Scripture regarding creation is thereby illustrated ; and that it is therefore manifest that it has
pleased the Almighty, in framing the world and all its
creatures, to work according to plans which indicate
design.

But we may carry this further. Notwithstanding
the anatomical resemblance between man and the lower
animals—on which alone this false theory has been

are great and significant,' p. 104. He speaks of the ' present enormous gulf' between the human and the ape races, p. 102, note ; and
of the ' immeasurable and practically infinite divergence of the Human
from the Simian Stirps,' p. 103, note.

† Ibid. p. 104.

built—there is, both in man and them, a marked relationship between their actual physical organization and their respective habits and dispositions, which is quite fatal to the idea of community of origin. Man, without those mental endowments which distinguish him from the brutes, would have stood a poor chance against them in the struggle for existence. Defencelessness in various ways is a characteristic of his organization, when compared with that of the creatures with whom he would have to contend. This defencelessness, however, which would place man at such a disadvantage had he no mental powers, is more than compensated for by the gifts of reason and invention; and this may be said of the earliest examples of men which have ever been found. The Duke of Argyll observes:—' In proportion as the difference between man and the lower animals is properly appreciated in the light of nature, in the same proportion will the difficulty increase of conceiving how the chasm could be passed by any process of transmutation or development. This difficulty is still further increased if we advert for a moment to the direction in which the human frame diverges from the structure of the brutes. It diverges in the direction of greater physical helplessness and weakness. That is to say, it is a divergence which, of all others, it is most impossible to ascribe to mere Natural Selection. The unclothed and unprotected condition of the human body, its comparative slowness of foot, the absence of teeth adapted for prehension or for defence, the same want of power for similar purposes in the hands and fingers, the bluntness of the sense of smell, such as to render it useless for the detection of prey which is concealed—all these are features which stand in strict and harmonious relation to the mental powers of Man. But apart from these, they would place him at an immense disadvantage in the struggle for existence. This, therefore, is not the direction in which the blind forces of Natural Selection could ever work.'* If we imagine man to be descended from a monkey, for example, the creature which by Natural Selection made the first move towards the future man, would stand a poor chance with his former companions to maintain his position, and even existence, in the struggle for life. This seems to be fatal to the theory here combated; for the tendency would be for the intruder to be overcome, and the old stock to be thus preserved unimpaired.

Mr. Darwin's *Descent of Man*, published during the present year, is a work overflowing with facts, put together in a very interesting manner; but with the untempered mortar of free speculation, so as to make his conclusions utterly untrustworthy in themselves, quite apart from their being in direct contradiction to the statements of Divine Revelation. All through the work is to be traced a tone of arbitrary assumption and hypothesis; but so buried in the torrent of facts, that the sanguine reader is carried securely along, though the logical mind is perpetually receiving a shock. Did we know with certainty from some independent source, that all living things have been really derived by inheritance from an original monad, Mr. Darwin's book, which shows such great research and ingenuity, would have produced this

* *Primeval Man*, pp. 65–67.

R

result:—It would have furnished us with some few scattered and fragmentary *illustrations* of the truth already known. I say 'few,' undoubtedly numerous as his facts are, in reference to the millions upon millions of lines of genealogical descent which must have been lost, and millions upon millions and millions beyond them of links which are missing in those lines of which present species are supposed to be the last. To reverse this process, however, and try to ascend from existing facts to the original monad, which is Mr. Darwin's aim, differs from the former as much as making a pyramid stand steadily upon its apex differs from balancing it on its base.

Mr. Darwin attempts to build his argument for the descent of man from some inferior ape-like animal upon a threefold basis: (1,) Similarity in the bodily structure of man and other mammals; (2,) Identity of character in the earliest stages of embryonic development; and (3,) The existence of rudimentary organs in man, apparently of no use to him, but of use in other animals. I will take up each of these points.

(1.) *Bodily structure.* Since the appearance of man in the world it has ever been obvious, that there is a certain resemblance between him and other mammals. They have two eyes, two ears, one nose, one mouth, four extremities (whether two arms and two legs, or four legs); and the points of resemblance have been greatly multiplied as knowledge has advanced. Comparisons, too, have been made between the habits of man and of the irrational creation, and men are said to be lion-hearted, chicken-hearted, cocky, henpecked, ravenous, eagle-eyed, ostrich-like, sheepish, pig-headed, swinish; a human being may be a dove, a lamb, a goose, a duck, a donkey, a snake in the grass, a bear, and a brute. These are terms and phrases perfectly familiar to us. But now to be gravely told that these creatures are our cousins, though it may be of the millionth, billionth, or trillionth degree, is somewhat astounding, and—requires proof. I am happy to say that no proof whatever is given in Mr. Darwin's book. His theory is simply based upon pure speculation, flowing from certain points of resemblance; and illustrates very forcibly the loose and presumptuous nature of much of the philosophizing of the present day. It appears to me to be a disgrace to the spirit of the age, that such speculations, in which the very point aimed at is assumed, should be seriously accepted as representing scientific truth. Mr. Darwin is obliged to avow, 'connecting links have not hitherto been discovered' (vol. I. p. 185); 'we have no record of the lines of descent, these lines can be discovered' —should he not have said guessed at?—'only by observing the degrees of resemblance between the beings which are to be classed' (p. 188)—all very good if it was known at starting that these lines exist; but altogether gratuitous and baseless in the absence of that *à priori* knowledge. He acknowledges that man's 'body is constructed on the same homologous plan as that of other mammals' (p. 185). Then why not be satisfied with this view? That it pleased the Almighty to make His creatures after a common plan, some parts being more developed in one class than in another, meets all the requirements; and, moreover, is not contradictory to the Scripture record. All that Mr. Darwin asserts, even of there being 'no fundamental difference between man and the higher mammals in their mental faculties' (p. 35)! would prove nothing for community of origin, were his assertion true ; but could only point out a certain similarity of character in those particulars.

Mr. Darwin replies, however, that 'it is no scientific explanation to assert that they [the hand of a man or monkey, the foot of a horse, the flipper of a seal, the wing of a bat] have all been formed on the same ideal plan' (pp. 31, 32). But perhaps here a 'scientific explanation' is out of place. Science is knowledge derived from an examination of the laws which regulate the physical world when once things are come into existence. It has nothing to say to the reason of the laws with which things were at first endowed. If it pleased God to create animals according to a plan, we have to accept this; and it is out of place to seek for a 'scientific explanation' of what it pleased the Almighty to do. Precisely the same objection might be urged to Mr. Darwin's own conception of the beginning of things as unscientific, viz. of 'life with its several powers having been originally breathed by the Creator into a few forms or into one.'* We must have a beginning. But Science is incapable of showing what it was ; it can only trace the phenomena of things as they are and were, after coming into existence. It is only He who made the world that can tell us of the beginning.

His theory has taken such hold upon him, that

* *Origin of Species*, 1869, p. 579.

he perseveres as if persuaded that it is absolutely true. 'We are quite ignorant,' he writes, 'at how rapid a rate organisms, whether high or low in the scale, may under favourable circumstances be modified' (p. 200) —in fact, he should have added, we know not whether they can be permanently modified at all. 'We know, however,' he goes on to say, 'that some have retained the same form during an enormous lapse of time'— and he should have said, that we have no evidence whatever that at the beginning of that enormous period any change occurred. 'The great break,' he continues, 'in the organic chain between man and his nearest allies, which cannot be bridged over by any extinct or living species, has often been advanced as a grave objection to the belief that man is descended from some lower form; but this objection will not appear of much weight'—to whom?—'to those who, convinced by general reasons, believe in the general principle of evolution' (p. 200)—that is, who take for granted the thing to be proved! 'With respect,' he says, 'to the absence of fossil remains serving to connect man with his ape-like progenitors no one will lay much stress on this fact, who will read Sir Charles Lyell's discussion in which he shows, that in all the vertebrate classes the discovery of fossil remains has been an extremely slow and fortuitous process' (p. 201) —and, therefore, I would reply, if we *knew* that the theory is true, we should be sanguine that some day proof would be found in fossils; but as the whole is a gratuitous hypothesis, the entire absence of fossil proof is a stern rebuke to the speculators. 'The belief,' he goes on to say, 'that animals so distinct as a monkey or elephant and a humming-bird, a snake, frog, and fish, &c., could all have sprung from the same parents, will appear monstrous to those who have not attended to the recent progress of natural history. For this belief implies the former existence of links closely binding together all these forms, now so utterly unlike' (p. 203). And he attempts to quiet us by saying, that some animals have been found, living or fossil, which lie between these species so as to help to bridge over the differences. But it is not units, or tens, or even hundreds of links, but millions, which we want to connect such different species; and then after all to be convinced that they are links.

His argument from the structure of man's body wants reality; whereas all his facts go forcibly to illustrate what has for ages been observed, that a certain plan, the design of an Almighty Intelligence, is to be traced throughout creation.

(2.) The second ground on which Mr. Darwin attempts to base his theory is *Embryonic Development*. 'No other explanation,' he says, 'has ever been given of the marvellous fact that the embryo of a man, dog, seal, bat, reptile, &c., can at first hardly be distinguished from each other' (p. 32). But I deny that Mr. Darwin's theory has given any explanation of this remarkable phenomenon; and further, I affirm, on the other hand, that it very remarkably establishes the opposite view, of creation having been according to a plan. For be it remembered that the embryo of a man is the offspring of a man and becomes a man and nothing else; the embryo of a dog is the offspring of a dog and becomes a dog and nothing else; and so of each example he produces. This has always been going on, and no departure from it has ever occurred, except in the imagination of Mr. Darwin and his followers. They have invented the change from one kind of creature to another to support the theory; but not one instance can they produce. This is their argument. These embryos are undistinguishable from each other at an early stage; and therefore have an identity of character. The creatures, then, to which they belong, must have a *common origin*, though it must be placed myriads of ages ago, as not the slightest change in this direction has been detected during all the years over which observation has ranged. But here a vast amount of fogginess exists and obscures the subject. Here is one of those leaps in the dark, a spring upwards, which obscurity so well conceals and assists. Why a common origin? Why carry up the imagination along these lines of life through myriads of ages back, hopelessly dreaming of an ultimate convergence of the lines? I say hopelessly, because (1) observation has never given the faintest indication of such convergence, that is, of divergence downwards, and (2) because in every individual now born you see the needlessness of such a search. You have not to go back myriads of years, but merely through a fraction of one year, to see, according to this theory, that each animal at its birth has recently come from an embryo identical in character with all the embryos of other animals in their earliest stage. So that, although these embryos are so precisely alike, they *rapidly*—not by slow gradations, taking ages for their work—pass into all the variety of animals we see around us. These lines are absolutely independent of each other, and what similarity there is in the creatures has arisen from causes which belong to them severally, and springs not in any way from community of being; for the causes have existed and operated within an interval of time, over the whole of which observation has watched their action and proved them to be absolutely distinct.

This supposed or apparent identity, however, affords a strong argument for the opposite view; that God has remarkably made the world after a plan; for here you have a perpetual recurrence of apparent identity, which cannot possibly be actual.

(3.) Mr. Darwin's third argument for his theory of the descent of man is laid in his assertion, that in the human frame there are useless parts in a rudimentary state, which exist in full use and development in other animals. 'In order to understand,' he says, 'the existence of rudimentary organs we have only to suppose, that a former progenitor possessed the parts in question in a perfect state, and that under changed habits of life they became greatly reduced' (p. 32). Thus man's immediate predecessor is said to be of the monkey kind; but we have to pass to another line, descending from some common progenitor, that in the horse, with its peculiar muscle, the 'panniculus carnosus' (p. 19) by which he can twitch his whole skin to perfection, we may see the hereditary principle on which we can raise the eye-brows, shift the ear, and move the scalp, which some men can do to a wonderful degree — a most luminous and convincing proof of our common descent! But why should we cloak our ignorance by asserting that the so-called rudimental organs in man are of no use? Would not modesty rather say, that their use, occupying as they seem to do in man an inferior position, is not yet ascertained?

The whole meets with a solution, as before, in the idea, that the Creator has worked upon a plan, and parts of that plan have been unfolded in one way in one set of creatures, and in another way in another; and that all parts, no doubt, have their several uses, though we have not discovered them in every instance. But Mr. Darwin takes a very different view. Not to admit this community of descent is, he asserts, to 'admit that our own structure and that of all animals around us is a mere snare laid to entrap our judgment' (p. 32)! I reply, No; but it strongly illustrates, that Divine wisdom was pleased to follow a type in creating His creatures. Mr. Darwin considers it to be 'only our natural prejudice . . . and arrogance . . . which leads us to demur to this conclusion,' that we are of common origin with the creatures round us. I reply, No: it is our thankful recognition of the Word of God, which has here spoken to us on a point, viz. the origin of things, which is above and beyond the province of Science to decide or even to indicate. Mr. Darwin considers that the facts his diligence has accumulated 'speak falsely,' if they speak not in confirmation of his humiliating and debasing theory. But, 'savage' as I may be in so doing (vol. II. p. 386), I still look upon all creatures as disconnected in their species by descent; but as uniting in their marvellous resemblances to testify to the wisdom of the Creator, who has bound them together in one harmonious whole, and watches every part with a Father's care, so that not a sparrow even is 'forgotten before God' (Luke, xii. 6).

The other part of this theory, so opposed to the Holy Scriptures, viz. that man, upon his first appearance upon the stage as man, was in a state of 'utter barbarism,' will stand scrutiny no better. Their argument is based upon this, that the habits of the earliest traces of man upon the earth were such as to betray an ignorance of mechanical arts; that traces of barbarous habits are found in civilized nations; that savage tribes are capable of improvement; and that no evidences of degradation are discernible. As to the first, why should a knowledge of the mechanical arts be arbitrarily chosen as a standard of primitive civilization, and the absence of that knowledge be taken as an evidence of barbarism? Do the angels understand the mechanical arts?—those lofty intelligences who excel in strength, who do God's commandments, hearkening unto the voice of His word? (Ps. ciii. 20.) Good order, mutual good-will and helpfulness, a sense of duty and corresponding obedience, are far higher indications of civilization than a knowledge and practice of all the arts in the world. 'No proof, if proof there be, that Primeval Man was ignorant of the industrious arts, can afford the smallest presumption that he was also ignorant of duty or ignorant of God. This is a fundamental objection to the whole of Sir J. Lubbock's argument.'* The same luminous writer, whom I have here and before quoted, rebuts the arguments of this theory at every point. Sir John Lubbock has replied to the Duke in the volume I have before

Sidenote: First man not a barbarian.

---

* *Primeval Man*, p. 132.

referred to, but in my opinion without success. He must be quite unaware of the gross ignorance and utter absence of religious feeling into which masses of people, even in close contact with the highest civilization, in London and some other great cities, can sink, when he denies that a tribe can lose a religion which their forefathers held. With regard to traces of barbarous habits found in civilized nations, the Duke says, 'Traces or remains of barbarism, properly so called, that is, traces of customs savage and immoral, in the usages of civilized nations . . . afford no presumption whatever that barbarism was the Primeval Condition of Man, any more than the traces of Feudalism in the laws of modern Europe prove that feudal principles were born with the human race. All such customs may have been, and as many think probably have been, not Primeval but Medieval, that is to say, the result of time and development, and that development a development of corruption.'* He then takes cannibalism, as an evident result of corruption—a loathsome practice which cannot, on Sir J. Lubbock's own admission, have been primeval. And why, then, may not other barbarous customs, found in more or less civilized nations, have sprung up in the same way, from the corruption of man's nature? Their existence in Primeval Man is no more to be believed than cannibalism itself. That there are indications of progress even among savages, in some respects, there may be no difficulty in allowing; but this in no way establishes the theory that there was not a previous degradation; it in no way makes it probable that the savage state

* *Primeval Man,* p. 133.

was the primeval state. Mr. Tylor, in his work on the *Early History of Mankind*, considers that the facts he has collected together favour the view, that the wide differences in the civilization and mental state of the various races of mankind are rather differences of development than of origin, rather of degree than of kind: and that in such practical matters as relate to the useful arts the history of mankind on the whole has been a history of progress. This seems to be clear. ' So far as may be judged,' he writes, ' from the scanty and defective evidence which has as yet been brought forward, I venture to think the most reasonable opinion to be, that the course of development of the lower civilization has been on the whole in a forward direction, though interfered with occasionally and locally by the results of degrading and destroying influences.' He acknowledges that there have been ' degrading and destroying influences,' and says there are no sufficient materials for the formation of a definite theory of the rise and progress of human civilization in primitive times.* Here he appears to speak more especially of civilization as displayed by a knowledge of the arts of life. For, as the Duke of Argyll says most truly,† ' nothing in the Natural History of Man can be more certain than that morally, and intellectually, and physically, he can, and often does, sink from a higher to a lower level. This is true of Man both collectively and individually—of men and of societies of men.' History and monu-

* Tylor's *Researches into the Early History of Mankind.* Second Edition, pp. 370, 372, 374.
† *Primeval Man,* p. 156.

ments of bygone people demonstrate this beyond dispute.

But there is a further point in this discussion. History and observation seem very clearly to teach us, that the most ancient remains of man hitherto discovered did not belong to the primeval members of the human family; but to outcasts and wanderers seeking a precarious existence far far from the primitive home of man. The Duke of Argyll points out that the most degraded and wretched tribes of man are now to be found at the extremities of continents and in the most inhospitable regions of the earth; as the Esquimaux in the North (to whom, however, Sir John Lubbock objects as proper illustrations) and the Patagonians in the South of the great continent of America; the Bushmen in South Africa; and the native inhabitants of Van Dieman's Land, 'the most utterly degraded of all the Polynesian races.' The natural increase of population, and the internal feuds and wars which ensued, would drive the weaker tribes to the less favourable regions, and the races driven furthest would become the rudest, the most engrossed in the pursuits of mere animal existence. 'And now we can better estimate,' our author proceeds, 'the value to be set on the arguments which have been founded on the rude implements found in the river-drifts and in the caves of modern Europe [and in other parts of the world]. . . . . It must be as safe to argue from those implements as to the condition of man at that time in the countries of his primeval home, as it would be in our own day to argue from the habits and arts of the Esquimaux as to the state of civilization in London or in Paris.*

11. The case which I have been just considering is the origin of man, in answer to those theorists who throw out the wild and groundless suggestion that he is descended from some inferior animal, and that in this line of descent 'the first beings worthy to be called men' were in a state of 'utter barbarism.' I will now produce another instance of their going counter to the teachings of Holy Scripture; and will show that here also nothing has been proved against the Sacred Volume. The Origin of Life—by which, I mean, the primitive conditions under which life appeared as regards the organism which it first possessed, and also the source from which life sprang to possess that organism—is a subject which has been engaging their attention now for some time. The microscope and incessant research, and a spirit of unrestrained speculation, have led to the results they announce. Both Mr. Darwin and Professor Huxley consider that life first appeared in the animal and vegetable world in some extremely simple primordial organism, of which all living things which exist now, and ever have been, are the lineal descendants. With regard to the source from which life came in this organism these two naturalists differ. Mr. Darwin, borrowing his idea, it would appear, from Scripture, believes† that the Almighty breathed life in the first instance into this supposed progenitor of the whole animal and vegetable kingdoms;

*Primeval Man, pp. 179, 180.
† Origin of Species. Second Edition, p. 484.

Marginal note: Origin of Life: its first forms.

whereas Professor Huxley traces life to the molecular forces, which are essential properties of the matter itself of which the monad consists. These views are utterly irreconcilable with what I have quoted at p. 227 from the book of Genesis. There the present vegetable kingdom first appears from the hand of the Creator as grass, herbs, and fruit-trees, all created with the seeds within themselves for future generations (Gen. i. 11, 12).* Moving creatures that have life in the waters, fowl flying in the air, cattle and creeping things and beasts of the earth, make their appearance, all after their kind, and finally man himself, all in the condition in which we see them now (v. 20–28). It may have suited Mr. Grove's purpose, when President of the British Association, and he may have caught the attention of a popular assembly to his views, by asking how they could conceive a full-grown elephant suddenly appearing upon the earth, and inquiring whence it could have come; could it have dropped down from heaven? But thinking men might have seen that there was no valid argument involved in this. What has greater and smaller to do with the question? The sudden appearance on the earth of the supposed monad, containing in itself such marvellous powers that it would become the parent of all living things through all time, would have been a wonder,—however far back these theorists place its appearance,—as great as the appearance of a full-grown elephant. Such modes of reasoning are utterly vain. This is a kind of clap-trap argument quite unworthy of a philosopher.

* See Note at p. 66, last paragraph.

If we investigate the order of nature as it exists, we see that the seed is necessary to the production of the tree, the embryonic germ to that of the animal; and also, on the other hand, the tree is necessary for the production of the seed, and the animal of the germ. The succession of life occurs in a perpetually recurring series, each point in the series, wherever we select it, being necessary for the existence of what follows it, according to the observed course of nature. We see the egg, the caterpillar, the chrysalis, the butterfly; the egg, the caterpillar, and so on. We may begin with any link, the chain is the same.

Even in their theories of development and evolution, in which permanent changes are supposed to have taken place after periods of enormous duration, each point in the succession is essential to all that follows it, wherever we select it. And there is no reason why their hypothetical monad, or even protoplasm, the simplest state of matter in which life has yet been traced, should not be regarded in reality as complicated in its structure as an elephant or any other existing creature. It seems, indeed, as if the advance of knowledge has the tendency of filling some of us with self-confidence rather than humility. They seem to think that they have penetrated backwards to a natural beginning of things. The microscope has added so greatly to man's natural powers of observation, and they have got so far back, that they think they have got to the origin and fountain-head. Here they think that they have reached a simple being, than which nothing can be simpler. But was not each creature, till physiology had become a science, regarded much in the same light? Physiologists have now investigated the wonderful works of God in creation, and learnt, that (to our apprehension) the simplest creature is a mere cell with life, and that all vegetable and animal beings are built up of these cells. And Professor Huxley would go further back and assert that we can trace the origin even beyond the cell, to protoplasm, which he regards as the first existent matter of life. But why should we stop here? Because our microscopes fail to guide us: their powers are exhausted. But are the secrets of nature exhausted? Why may there not be in the minutest speck which our microscopes can detect worlds within worlds of organisms? The minutest speck is infinitely greater than nothing. It has its two halves, and each half has its two halves, and so on without limit, and the matter can never thus be exhausted. We can never get to a point where we can say, Beyond this all complexity of organization absolutely ceases: Here we reach the absolute *element* to which life clings. Such reasoners forget the beautiful thought of Chalmers, that no sooner was the principle of the telescope discovered which unfolds to our view worlds beyond worlds in the immensity of space, than the same principle gave us the microscope, to show us the unlimited extent of God's works in the opposite direction; and that while our finite powers may set a limit to our observation, no limit can be fixed by us either to immeasurable magnitude in the one case, or infinitesimal minuteness in the other. The phenomena of light give us indisputable proof, that even our own eyes are susceptible of impression from minute portions of matter

s

incalculably smaller than the smallest speck they can actually see with the most powerful aids, and are thus cognizant of their existence. For they can at once without that aid discern the difference, for instance, between the colours of violet and blue in the spectrum : and yet that difference depends upon a minute quantity which equals only the millionth part of an inch ; that is, the difference between the length of a wave of violet and one of blue light, a quantity too small to be seen by microscopes of the highest power ; and further still, even those waves, and therefore their difference, are made up of innumerable parts, under all degrees of condensation and rarefaction, all having their effect upon the eye in producing the impression of the respective colours. If, then, our limited faculties can be cognizant of the existence of such unspeakably minute objects which the eye cannot see even by the help of microscopes, and which the imagination alone can realize, though really existent, what a vast unknown field there is beyond that in which the phenomena of protoplasm are detected by the keenest aided-vision, in which operations as wonderful, and perhaps still more so, are going on ! As regards organization, then, we have no ground for supposing that in protoplasm we have arrived at absolute simplicity, and have discovered matter in such a state, that life is its natural property. We cannot pronounce that the structure in any one stage is really less complicated than in another. The first link in the chain of being, in the theories of development and evolution, may be, as far as it is a living organism, as complex as any link which follows ; and one might say, even more so, as it contains within itself all creatures which flow from it by natural descent.

In the chain of succession, then, of species now existing, or in the imaginary chain in the fossil world, there is nothing in Science to indicate at what point of the sequence creation took place. With regard to species now existing, the statement of Scripture is clear, that they were brought into the world by the fiat of the Almighty as grass and herbs and trees, with the seed within themselves, and as fish and fowl, cattle and beasts, also possessing powers of reproduction. Here, then, Science proves nothing contradictory to the declaration of Scripture regarding the organism which life first possessed, and they are not at variance in this matter; although it may appear to some, who prefer their own speculations to the word of God, that this development theory gives a nobler view of the works of God than God has given Himself!

But I have more to say on this theory. Let us consider the lines of descent from their hypothetical monad and beginning of life. On their own showing * these naturalists can produce no instance of permanent change in any species now existing, all through the generations after generations which have been born and have died in each of these species. When we ascend these lines and come to their supposed continuations backwards among the fossil records of the earth, what do we find? A continuous series of links? Nothing of the kind. Only isolated and widely separated links, like dots, wide apart in space, grouped

* See Professor Huxley's Address as President of the Geological Society, 1870, reported in *Nature*, Feb. 24, 1870.

into lines solely from some resemblance, and altogether wanting (on their own acknowledgment), it is not known how many, it may be millions, of intervening links, which can be nowhere found, to fill up the succession of countless hiatuses which yawn along the lines. I can hardly see any more reality in the supposed relationship of these scattered species of organism in the fossil world, than I can in the existence of Orion, the great and mighty hunter, in the heavens, because the Greeks arbitrarily grouped together a number of isolated stars and traced the giant's imaginary form. These naturalists go solely by resemblance. They cannot give one proof of actual descent of one species from another species in the fossil world, and when they come to the existing world of living creatures, and observe their laws and watch their birth, no change of species whatever can be detected. With regard to the fossil world they reply, that the record is imperfect. But strange and, indeed, incredible is it, that only single isolated links should have been left; and that when hundreds of intervening links have disappeared, sometimes *two* links in contact, to say the very least, should not have been left remaining to tell their tale! Attempts and guesses have been made to discover some true linear types. Professor Huxley says in his Address to the Geological Society, in 1870, 'It is no easy matter to find clear and unmistakeable evidence of filiation among fossil animals. For, in order that such evidence should be quite satisfactory, it is necessary that we should be acquainted with all the most important features of the organization of the animals which are supposed to be thus related; and not merely with the fragments upon which the genera and species of the palæontologist are so often based.' And after noticing some supposed examples of linear types produced by M. Gaudry and Professor Rütimeyer, he adds :—' But, as no one is better aware than these two learned, acute, and philosophical biologists, all such arrangements must be regarded as provisional, except in those cases in which, by a fortunate accident, large series of remains are obtainable from a thick and wide-spread series of deposits. It is easy to accumulate probabilities—hard to make out some particular case in such a way that it will stand rigorous criticism.' He then adduces a case which he thinks can be thus made out in favour of the pedigree of the horse. But it is clear to me that his leading idea, in itself an assumption, influences his judgment and induces him too easily to turn resemblances into proofs. The genus Anchitherium appears in the upper eocene and the lowest miocene. In the middle miocene appear the two genera, Hipparion and Anchitherium. In the upper miocene appears the genus Equus, which from certain resemblances is *assumed* to have grown out of the other two in succession. But how natural selection can have changed a large part of a *genus* it is not easy to see. Natural selection would affect individuals, and not even whole species, since birth or origin runs in lines of individuals. The only creatures which have any influence upon the nature of an animal are its two parents; other members of the species, much less of the genus, have no effect whatever upon it. Hence, if divergence does take place by natural selection, the shades of resemblance would have been countless, and of gradations almost imperceptible. The differences pointed out by Professor Huxley, which distinguish the genera Anchitherium, Hipparion, and Equus, are too great for his purpose, in the absence of many links. If he *knew* that his theory of evolution was true, then these three genera might be links of one chain, even though we miss the connecting links. But, since the conjecture regarding evolution is the thing which has to be proved true, the absence of these links, even in such a case as the present, where certain resemblances exist, is a fatal stumbling-block to any but to sanguine minds which seem ardently to desire *their* theory to be the law of nature. This theory of development presents this extraordinary phenomenon, fatal to any theory in the eyes of men of sound philosophical perception, viz., the assumption of the very thing which has to be proved.

I would add, that the method which these naturalists pursue is the deductive method. They assume a principle, and compare its consequences with facts, as far as they can. It may be said, indeed, that the theory of gravitation is really established by this deductive process. That theory, however, was first attained by induction from facts ; viz. those chiefly which Kepler had by incredible diligence gathered from an examination of the heavens, and comprised in his three celebrated laws. Newton by the first of those results proved that the force acting on all the planets, whatever it was, must pass through the sun ; by the second, that the force on each, whatever its intensity, must vary as the inverse square of the distance of the planet from the sun ; by the third, that the actual amount or intensity of force (estimated at a given distance) must be absolutely the same for all the planets. That is, that one and the same force resides in the sun (varying in its effect as the inverse square of the distance), and regulates the motion of the system. This law was thus arrived at by pure induction from facts. Its final confirmation, however, has been obtained by the deductive process; by assuming its truth, and comparing its effects with the facts of nature in innumerable observations and experiments spread over two centuries.

But suppose even, for the sake of argument, that the law of gravitation was in the first instance assumed to be true, as this law of development is, and proved to be true by its accurately explaining complicated phenomena; there is this very marked difference between them. In the application of the law of gravitation each phenomenon is brought into direct connexion with the law itself: no connexion of the phenomena one with another is necessary, but each is shown to be a direct deduction from the law. Not so in this theory of development. The very soul of this theory is the possibility of showing a relationship between the phenomena themselves. Here is their great difficulty; and I think we cannot but see, in reading Professor Huxley's Address to the Geological Society, just referred to, where he dwelt upon this subject, and attempted (in the case of the horse) to connect phenomena with each other, how the controlling idea of his assumed law influences his decisions where facts alone ought to have satisfied him. There are many facts in his interesting address, but many bridges created by his magic wand, like castles in the air, to connect the facts together, which vanish upon a touch, and leave the hard facts as isolated as before. The untiring research upon these topics which able and industrious men are engaged in is adding vastly to our knowledge. This knowledge, which we owe to their diligence, is not lost, although their speculations may be groundless.* The late Mr. Hopkins has well remarked, that we sometimes meet with 'the union of great acuteness of observation and large views in the grouping and classification of natural objects and phenomena, with the want of that judicial power by which a man is enabled to estimate the true weight of the evidence laid before him and the degree of confidence with which it ought to inspire him.' †

But now, regarding the source from which life comes. Mr. Darwin traces life itself, in his first monad, Origin of Life: its Sources. up to the Almighty, as the direct bestower of it. Not so Professor Huxley. He has no doubt distinctly avowed, in his recent Address as President of the British Association at Liverpool, that observation and experiment fail to detect any other source of life in things now brought into existence than previously existing living things. *Omne vivum ex vivo*,—a principle which he traces to the Italian naturalist, Francesco Redi, of the seventeenth

---

* 'If they [the theoretical views of the '*Origin of Species*'] were disproved to-morrow, the book would still be the best of its kind — the most compendious statement of well-sifted facts bearing on the doctrine of species that has ever appeared.'—Huxley's *Lay Sermons, &c.*, p. 327.

† See two admirable papers in *Fraser's Magazine*, June and July 1860, by Mr. W. Hopkins, on 'Physical Theories of the Phenomena of Life'—a careful perusal of which I earnestly commend to all inexact speculatists.

century,—he maintains is a principle which facts teach us now on the origin of life. But he nevertheless, in the same address, declares his philosophic 'expectation,'* that had he seen the earth when passing through its earliest physical and chemical conditions, previous to all geological time, he would have witnessed 'the evolution of living protoplasm from not living matter.' Nor does he appear to give up all hope of this evolution being witnessed at some future time. He retracts nothing, in fact, from the statement he has made on this theory of evolution in his previously published works. The speculations therein unfolded have led him to these conclusions: that a certain substance, which he calls protoplasm, lies at the basis of all vegetable and animal life; he calls it 'the matter of life:' that plants and animals present this difference; plants can manufacture fresh protoplasm out of mineral compounds; whereas animals are obliged to procure it ready made, and hence, in the long run, depend upon plants: that protoplasm contains the four elements, carbon, hydrogen, oxygen, and nitrogen, in very complex union, viz., as carbonic acid, water, and ammonia: that as hydrogen and oxygen, after an electric spark passes through them when mixed in certain proportions, become water, an apparently new substance with certain pro-

* He calls it also 'an act of philosophic faith,' confounding expectation with faith, which is belief *on testimony*, the very thing which is here altogether wanting. Bishop Pearson has long ago told us that belief rests upon three distinct grounds, and has accordingly three distinct names. I may believe, because I know; that is science: or because of the balance of probabilities; that is opinion: or because I am told; that is faith.

perties; so the four elements mentioned above, when brought together in certain proportions and under certain circumstances (what these circumstances are is the question), produce protoplasm, which, as the case may be, exhibits all the phenomena of vegetable or animal life: and that as we do not attribute the properties which water manifests to a principle, which, for the sake of a name, might be called 'aquosity,' no more are we at liberty to attribute the properties which protoplasm manifests to a principle which we call 'vitality:' and he comes to this conclusion, 'that all vital action may be said to be the result of the molecular forces of the protoplasm which displays it.' And in addressing his audience, he adds, 'If so, it must be true, in the same sense, and to the same extent, that the thoughts to which I am now giving utterance, and your thoughts regarding them, are the expression of molecular changes in that matter of life which is the source of other vital phenomena.'[*] How this differs from materialism I find it impossible to see. And yet the author adds, 'I am no materialist, but, on the contrary, believe materialism to involve grave philosophical error.'[†] At any rate, whatever name he may choose to give to his protoplasmic idea, there is a very obvious difference between the results of his ingenious principle 'aquosity,' and of our old-fashioned principle 'vitality.' In the first place, the parallel does not hold. Had the ugly word 'protoplasmity' been used, and not 'vitality,' the case would have been

---

[*] *Lay Sermons, Addresses, and Reviews,* by Professor Huxley, VII. 'On the Physical Basis of Life,' pp. 132, 138, 143, 149, 150, 152.
[†] *Ibid.* p. 152.

different. The new word 'aquosity' is a mere synonym, derived from Latin, of what belongs to water: it means the properties* of water, and is therefore necessarily associated with water. But vitality is not necessarily associated with protoplasm, as Professor Huxley admits that there is dead protoplasm as well as living. The very fact that protoplasm can exist in the two states, alive and dead, is a sufficient proof that there is some such principle as vitality, or life, which is separable from protoplasm, and is not an essential property of it. Again: though water and protoplasm have this in common, that the elements which go to form them must be acted upon by an external agent, and will not unite when merely placed together; though they have no inherent power of forming either water or protoplasm of themselves; yet how widely different are the agents and the results they bring about. When hydrogen and oxygen are united by the electric spark, the product water remains water for ever; till a change takes place in its temperature, when it may become steam or ice in any of its forms; and will return to water when the temperature is again altered. But how different is the effect of vitality in forming protoplasm out of the 'lifeless' elements.† It builds up from it plants and animals, which are not stationary, like water, but are ever growing and enlarging, and throwing out varied forms. Once more. In none of the states which water assumes does its 'aquosity'

---

* On the limited applicability of this term 'properties' to inanimate matter, see a valuable article on 'The Physical Basis of Life,' in the *Contemporary Review*, June, 1869, Professor John Young.

† *Lay Sermons, &c.* p. 149.

give it the power of *reproduction*. But the products of living protoplasm, endowed with this mysterious principle of 'vitality,' do possess that remarkable power, a power which at once indicates the presence of some principle totally distinct from anything like mere aquosity, and which can bear no analogy to it. It far transcends it, and, indeed, is a principle of a totally different order, inasmuch as it is self-acting and leads to an endless line of beings of the same kind, and increasing in numbers as successive generations are produced.

What, then, do we learn from the remarkable revelations which the microscope unfolds to our view? No more than this. What minute and (apparently) simple organisms it has pleased Almighty God to bring under the influence of life! But, as to the source of this mysterious principle, Science teaches us absolutely nothing, though it demonstrates its existence.

Interesting as these investigations upon protoplasm are, it must be remembered that it has long been known that all material things, animate and inanimate, are composed of one or more of certain definite elementary substances. The point brought to light by these more recent discoveries is, which of these elements are used; and it is shown that the same compound of these elements is at the basis of both the animal and vegetable kingdoms. But, indebted as we are to the naturalists who have demonstrated this to us, both for their unwearied diligence and the skill applied in a thousand ways for bringing these facts to light, I think that there is great danger of our over-estimating the greatness, or the extent, of the discoveries. This I say in no spirit of disparagement whatever of what they have done. By this, so to say, hacking and hewing and mangling of the delicate textures of animal and vegetable beings it is discovered, that, on reaching a certain way down, we come to the same living material, protoplasm, out of which all organic creatures, animal and vegetable, are built. Living plants have the property of appropriating to themselves certain lifeless chemical compounds and thus making protoplasm and growing thereby; which animals have not the power of doing. They must eat vegetables, or other animals which have been thus reared, or both, to maintain their existence. This is a beautiful theory. But when we say animals and plants are thus built up of protoplasm, what does this building-up mean? Certainly not the mere accumulation of heaps of this living material. What can Science yet show regarding the countless modes of action of the vital force upon protoplasm, so as in the vegetable kingdom to produce all the exquisite variety of forms and hues of flowers and trees, all the countless specimens of delicious fruits; and, in the animal kingdom, the varied forms and descriptions of fish, the beautiful plumage and song of birds, the elegance of the antelope and the horse, the nobleness of the lion, the grandeur and docility of the elephant, the face divine of man, and all the marvels of his frame, 'fearfully and wonderfully made?' Absolutely nothing. If this discovery be true, that protoplasm is a common basis of all living beings, it nevertheless sinks into insignificance in presence of what remains as yet uninterpreted and unintelligible, the interminable and varied modes in which vitality acts in producing all the beauty and variety of living objects out of the same material. Of this Science can teach us nothing whatever, but can only collect and classify the results.

As to the origin and source of life, these investigations throw no light upon the mysterious subject. It is unsatisfactory to see an attempt made to carry the subject into the dreamy region of metaphysics, and to confound matter and mind. Professor Huxley, in his paper on the *Physical Basis of Life*, says, 'Matter may be regarded as a form of thought, thought may be regarded as a property of matter,' and so hopes to justify in some way his hypothesis of the original evolution of life, and even thought, out of lifeless materials! And Professor Tyndall, in a lecture at Liverpool, indulges the hope, that not only the sun with its system, but that all life, all intellect even, and genius, may at length be traced to a primitive fiery cloud, from which all things animate and inanimate, material and spiritual, have been developed by evolution! Surely these men have lost their guide, the reason God has endowed them with; and, intoxicated with the very partial knowledge their investigations have brought to them out of the inexhaustible stores of which God's creation is full, appear to dream that they could have pointed out how the world should be created, and could have laid out its laws, if they had had but the opportunity. Rather than bewilder themselves in speculations utterly beyond their grasp, they had better rest contented with the truth contained in the well-known witticism, and learn the lesson it teaches of the uselessness, because hopelessness, of getting beyond it:—' What is matter? Never mind. What is mind? No matter.' In every line of investigation into natural things, our only sure method is to seize upon and demonstrate some intermediate fact, and from that as a starting-point proceed both ways, if we can, along the line. But we can never get to the beginning or the end of the line. Thus, that every particle of matter attracts every other particle of matter with a force varying as the mass of the attracting particle and the inverse square of the distance of the particles, is a fact in nature demonstrated now beyond question. And, by starting from this fact, Newton explained the complicated phenomena of the solar system. In the other direction nothing whatever has yet been discovered. No success whatever has been achieved in attempting to find the cause of this law of gravitation. So in the investigation of the phenomena of light, this is ascertained by innumerable experiments to be a fact of nature beyond all dispute ; that when a ray of light falls slanting upon the surface of a transparent medium it is bent on entering it according to a fixed and known law (called the law of the constant ratio of the sines). From this fact we proceed by the help of geometry (and other facts) to explain a large variety of phenomena, and to calculate the necessary forms of optical instruments. In this example, unlike the case of gravity, some progress has been made in the opposite direction. The phenomenon of light has been proved to be produced by the vibration of a highly elastic medium called ether. And (as is of course necessary to the proof of the validity of this hypothesis) the constant law of the sines referred to above is shown to be a necessary consequence of the existence of this ether. Here, then, we have been able to move some way in both directions from our central fact. But the central fact is still an immovable rock. With regard, then, to this question as to 'What is matter?' we have a central fact from which we start, and all the conclusions in one direction are indubitable truths; however cloudy, and even dark, the path may be, and proves to be, in the other or opposite direction. When I speak and reason about matter, I speak and reason upon what I feel, smell, taste, hear, and see. And whatever metaphysicians may say and dream, regarding these acts as acts of consciousness—which is unfathomably beyond their research—the conclusions drawn in the opposite direction regarding the phenomena of the physical world are altogether unaffected. For the naturalist, therefore, to carry back his statements and thoughts on natural things one step into the dreamy regions of metaphysics, dreamy because utterly beyond his depth, and where only his blind fancy and uninstructed imagination are the guides he can follow, is the height of imprudence.

If we give the reins to our fancy, we may be led into every kind of absurdity. From the things Professor Huxley has told us of protoplasm, we might, for instance, be tempted to suppose, had we not facts before us to contradict such a conclusion, that, as protoplasm is the material with which our frames are built up, and as this protoplasm is produced by vegetables only, and not by man or any other animal; man is the offspring of the vegetable kingdom, and there would be a literal illustration of the Scripture statement, 'All flesh is grass.' But happily, facts are on our side. We all know, that it is only when protoplasm has been built up in the form of men and women, that other men and women can be produced, whatever speculative theory might say; it is only when protoplasm takes the form of tigers, that tigers are produced; or of sheep, that sheep are brought forth. If all organic creatures are composed of this protoplasm in the way which these researches are supposed to indicate, then *this* is the inference we should draw:—How marvellously has the Almighty framed the various parts of His beautiful creation after a plan, and according to design; so that all creatures which have the gift of propagating their kind and perpetuating their race, however they may differ in their mature state, are built up, as far as we can see, from similar elements under the controlling action of a mysterious principle which He has implanted--the principle of 'life,' in both its kinds, animal and vegetable. As to the *origin* of life nothing whatever is discovered. *How* it was communicated, *when* it was communicated: regarding these Science is equally ignorant. In the course of nature the adult is necessary to the germ, and the germ to the adult; the tree to the seed, and the seed to the tree. In which of these conditions life first took possession Science is utterly unable to inform us: and therefore it teaches nothing on this subject contradictory to Scripture, which reveals to us that God created grass, herbs, and trees; fish, fowl, cattle, creeping things, and Man, all after their kinds. That was the commencement of life; and Science can produce nothing opposed to this. God endowed them with the gift of reproduction, which perpetuates their species down to our own time. The marvellous works of God are unfolded to our view by

Science. Scripture and Science are in no instance proved to be at variance by any of the theories I have been considering; but the mischievous ease with which some men can build up their hypotheses is painfully illustrated. What indeed are hydrogen and oxygen and nitrogen and carbon, but some of the materials of which the ground we tread upon is made? Science, then, in this instance bears its unequivocal testimony to the truth of that ancient inspired declaration, 'The Lord God formed man of the dust of the ground; and breathed into his nostrils the breath of life; and man became a living soul' (Gen. ii. 7): and the divine decree is perpetually receiving its fulfilment, 'In the sweat of thy face shalt thou eat bread, till thou return unto the ground: for out of it wast thou taken: for dust thou art, and unto dust shalt thou return' (Gen. iii. 19). All creation bears witness to this; that when the Almighty withdraws this principle of Life even in things irrational and inanimate, they are resolved into the dust of the ground; and when He gives life, these base materials assume the most beautiful forms. 'All flesh is grass, and all the goodliness thereof is as the flower of the field: the grass withereth, the flower fadeth: because the Spirit of the Lord bloweth upon it, surely the people is grass. The grass withereth, the flower fadeth; but the Word of our God shall stand for ever.' (Isaiah, xl. 6, 8.) 'Thou takest away their breath, they die, and return to their dust. Thou sendest forth Thy Spirit, they are created, and Thou renewest the face of the earth.' (Psalm civ. 29, 30).*

* I return to Professor Huxley's investigations in my 13th example on Design.

12. Another illustration of Science, under the guidance of unsound reasoners, appearing for a time, to some at least, to be at variance with the Scripture, is seen in the controversy regarding the Uniformity of Nature. It is the work of Science to gather together the phenomena of the material world, and to deduce from them the laws by which it is governed. The result has been the discovery that matter preserves its properties and laws with remarkable uniformity. Not a single case of departure from this uniformity has been detected during the whole period of modern science. The consequence is, that some have formed the belief that the course of nature is absolutely immutable; and that no amount of evidence can establish the fact of a departure from this constancy. Any new extraordinary phenomenon hitherto unexplained they would, under any circumstances, attribute to the operation of some law not yet known, so persuaded are they that the course of nature is absolutely immutable.

The Uniformity of Nature.

'In an age,' one writer says, ' of physical research like the present, all highly cultivated minds and duly advanced intellects have imbibed, more or less, the lessons of the inductive philosophy, and have at least in some measure learned to appreciate the grand foundation conception of Universal Law—to recognise the impossibility even of *any two material atoms* subsisting together without a determinate relation—of any action of the one on the other, whether of equilibrium or motion, without reference to a physical cause—of any modification whatsoever in the existing conditions of material agents, unless through the invariable operation of a series of eternally impressed consequences, following in some necessary chain of orderly connexion—however imperfectly known to us.'* 'The fact is,' another writes, 'that one idea is now emerging into supremacy in science, a supremacy which it never possessed before, and for which it still has to fight a battle; and that is, the idea of law. . . . The steady march of Science has now reached the point when men are tempted, or rather compelled, to jump at once to a universal conclusion; all analogy points one way and none another. And the student of science is learning to look upon fixed laws as universal.' But then follow these startling announcements :—'Many of the old arguments which science once supplied to religion are in consequence rapidly disappearing. How strikingly altered is our view from that of a few centuries ago is shown by the fact, that the miracles recorded in the Bible, which once were looked on as the bulwarks of the faith, are now felt by very many to be difficulties in their way; and commentators endeavour to represent them not as mere interferences with the laws of nature, but as the natural action of still higher laws, belonging to a world whose phenomena are only half revealed to us.'†

The late Mr. Buckle, too, in his *History of Civilization*, advances the same views; so much so as to state his belief that moral actions even come under some inevitable law, over which the actor has no control!—a result he comes to solely on statistical grounds, and

* *Essays and Reviews*, p. 133.
† Dr. Temple's *Sermon at Oxford, before the British Association for the Advancement of Science*.

those very imperfect, and utterly inadequate to support such a monstrous conclusion.*

This school even goes so far as to 'reject the idea of 'creation.'† If these be the teachings of Science, we have indeed a most signal instance of Science being at variance with Scripture; as Scripture declares that God created the world and all things in it, and in the miracles of the Old and New Testaments records many examples of departure from the ordinary course of nature carried out by human agency, though the events were superhuman and could proceed only from divine power. But an examination of this apparent discre-

---

* 'Suicide,' he gathers from his supposed laws of necessity, 'is merely the product of the general condition of society, and the individual felon only carries into effect what is a necessary consequence of preceding circumstances. In a given state of society, a certain number of persons must put an end to their own life.'—(Vol. i. p. 26.) Again, 'Such is some, and only some, of the evidence we now possess respecting the regularity with which, in the same states of society, the same crimes are necessarily reproduced.'—(p. 27.) 'Nor is it merely the crimes of men which are marked by this uniformity of sequence. Even the number of marriages annually contracted is determined, not by the temper and wishes of individuals, but by large general facts, over which individuals can exercise no general authority. It is now known that marriages bear a fixed and definite relation to the price of corn; and in England the experience of a century has proved that, instead of having any connexion with personal feelings, they are simply regulated by the average earnings of a great mass of the people.'—(p. 29.) This last illustration I have quoted, because I think it puts into our hands the means of disproving this dangerous theory. If the number of lawful marriages is proportional to the means of support—as the statistics are said to show—does not this demonstrate rather that a moral principle is at work, that people who live as they should and do not indulge in unlawful connexions, are driven by no law of necessity, but are influenced by right motives? To such extremes are the advocates of this theory of the Supremacy of Law driven by their speculations.

† *Essays and Reviews*, p. 129.

pancy leads to the same result as before, that the teachings of Scripture and Science, when properly considered, are by no means at variance.

(1.) In the first place, Scripture testifies to the general uniformity of nature as strongly as Science does. To mention only one instance: the Almighty appeals to that uniformity when assuring His people of His unfailing mercy. 'Thus saith the Lord, which giveth the sun for a light by day, and the ordinances of the moon and of the stars for a light by night, which divideth the sea when the waves thereof roar; the Lord of hosts is His name: If these ordinances depart from me, saith the Lord, then the seed of Israel also shall cease from being a nation before me for ever. . . . If ye can break my covenant of the day, and my covenant of the night, and that there should not be day and night in their season, then may also my covenant be broken with David.' (Jer. xxxi. 35, 36; xxxiii. 20, 21.) To this other examples might be added.*

(2.) Next, Scripture and Observation equally teach us that the physical world is not abandoned to the sole government of physical laws; but that moral and spiritual agency has a considerable influence upon their operation. If the world might be regarded as a complicated machine of merely material agents, there might be some reason in the expectation that its movements should be fixed in an inflexible succession. But such an idea the most zealous upholders of physical-law theories cannot maintain. Even human agency alone, as daily observation teaches, is continually breaking off

---

* See also Gen. viii. 22; Ps. lxxxix. 37, civ. 19-23, cxix. 89-91, cxlviii. 1-6; Isa. liv. 9, 10.

the action of one law, and that abruptly, and calling into operation another, which has hitherto remained unused, though ready for action. Two men equally bent upon self-destruction spring from a precipice. One, descending by the law of gravity, is dashed to pieces on the rocks below; the other in his downward course—seized with terror, or arrested by the voice of an imploring friend—grasps suddenly a suspended rope which the other passed unheeded by. His fall is checked; and hand over hand he mounts, regains his footing, and is saved! What has led to so marvellous a difference in the issue, when in the commencement the circumstances were so precisely alike? To a spectator the process, in either case, is seen to be the result of well-known physical laws: gravitation causing the descent; collision on the rock, death; the mechanism of the hand, the grasp; friction, the steady hold; the tension of the rope, the sustaining force; the muscular action of the arms, the re-ascent. The answer is this:— in the latter case a new element is introduced. MIND is suddenly brought into operation; and laws, ready for use, but not yet in action, are brought into play at a moment's call. We see this in all the events of life. Two men lie sick of the same deadly malady. The one abandons himself to the uninterrupted action of its physical laws, and dies; the other, prevailed upon by entreaty, calls in the physician, who brings other laws to bear, and the patient lives. It is strange that the writer I have quoted higher up from *Essays and Reviews*, seems to have made no allowance whatever for this obvious and ever-acting source of interruption to the unvarying operation of physical laws. In quoting from Archbishop Trench the following passage: 'We continually behold lower laws held in restraint by higher, mechanic by dynamic, chemical by vital, physical by moral'—he objects that 'the meaning of moral laws controlling physical is not very clear.'* But here, as my readers will see, lies the gist of my argument and illustrations.

In the same way Scripture teaches us that the Almighty in His providence is perpetually breaking off the action of one law and bringing another into play, and often in answer to prayer. When the Son of God Himself listened to the cry of His disciples, 'Lord, save us, we perish!' and 'rebuked the winds and the sea, and there was a great calm' (Matt. viii. 25, 26), though this was no miracle, or event contrary to the established course of nature, like the raising the dead to life, or feeding the multitudes and sending them away filled, with food sufficient only for a few men's repast—it was an abrupt change in the order of nature which would not have occurred but for His interposition: it was the audible exhibition of what the Almighty is perpetually doing, in ordering all things as it pleases Him, and often in answer to the supplications of His people. In this way the following fallacious style of reasoning may be met. 'A miracle is strictly defined,' Professor Tyndall says, 'as an invasion of the law of the conservation of energy. To create or annihilate matter would be deemed on all hands a miracle. The creation or annihilation of energy would be equally a miracle to those who understand the principle of conservation. Hence arises the scepticism of scientific

* *Essays and Reviews*, p. 133.

men when called upon to join in national prayer for changes in the economy of nature. Those who devise such prayers admit that the age of miracles is past, and in the same breath they petition for the performance of miracles. They ask for fair weather and for rain; but they do not ask that water may flow up hill; while the man of science clearly sees that the granting of the one petition would be just as much an infringement of the law of conservation as the granting of the other. Holding this law to be permanent, he prays for neither. But this does not close his eyes to the fact, that while prayer is thus impotent in external nature, it may react with beneficent power upon the human mind. That prayer produces its effect, benign or otherwise, upon him who prays is not only as indubitable as the law of conservation itself, but will probably be found to illustrate that law in its ultimate expansions. And if our spiritual authorities could only devise a form in which the heart might express itself without putting the intellect to shame, they might utilize a power which they now waste, and make prayer, instead of a butt to the scorner, the potent inner supplement of noble outward life.'* To this I would reply, that, whether or not mechanical energy is now created or annihilated, it is most certainly perpetually being diverted into new channels by the mind and skill of man, so as to produce results which would not otherwise have occurred. The whole of this argument, and all arguments like it, are based upon the tacit assumption that the physical world is abandoned to the sole operation of physical laws; which is absolutely false, as I show above. Such reasoning

* *Fortnightly Review*, December 1, 1865.

amounts to banishing the Almighty from His own world: for if the mind of man has its influence upon the course of nature, how much more may the Divine Mind have its influence, and that in answer to prayer. If the Almighty ever since the creation has seen good, for the benefit of His creatures, to govern the world by constant laws, has He therefore left the world? Cannot *He* call these laws into action or break off their action, if *man* can do so in his way? And more than this, cannot He even reverse their action if He sees right to do so, as in raising the dead?

In this connexion I may allude to the AGENCY OF ANGELS, as administrators of the Divine will in the present natural world, as repeatedly taught us in Scripture. 'He maketh his angels winds, and his ministers a flame of fire.' (Ps. civ. 4.) Kingdoms are committed to the superintendence of angels, as 'Michael the great prince,' and the angelic princes of Persia and Greece (Dan. x. 13, 20, 21; xii. 1):* the natural elements (Rev. viii. 7-12, &c. &c.) and individuals too (Matt. xviii. 10) are committed to angels. The visitations of the Almighty, whether in wrath or mercy, are often carried out through the ministration of angels. We need but mention the destruction of Sodom and Gomorrah (Gen. xix. 1-25); the rebuke of Balaam (Num. xxii. 22); the destroying pestilence in the days of David (1 Chron. xxi. 12-15); the destruction of Herod (Acts, xii. 23); the vials of God's wrath (Rev. xvi. 1). Then, also, the relief of Hagar (Gen. xvi. 7); the encouragement of Jacob (Gen. xxxii. 1, 2); the

---

* See also Deut. xxxii. 8; in the LXX. version of which the word 'angels' is found in the place of 'children of Israel.'

safety of Daniel (Dan. vi. 22); the rescue of Peter (Acts, v. 19); strength communicated from heaven to our suffering Redeemer (Luke, xxii. 43)—are all administered by these messengers from the courts above. In most, if not all, of these cases, the angelic agency manifests itself by the use of the constituted laws of nature. When we maintain that the Almighty for His own purposes can change the ordinary course of nature, and has done it, when He has seen fit, in miracles, we are not bound to assume that He actually violates any of the laws He has impressed upon the world, material and spiritual; but that He works by them, in a superhuman, but not necessarily supernatural way. He can create, and He can annihilate. He can do all things. But He appears to rule by law. 'The language of Scripture,' says the Duke of Argyll, 'nowhere draws, or seems even conscious of, the distinction which modern philosophy draws so sharply between the natural and the supernatural. All the operations of Nature are spoken of as the operations of the Divine Mind.'* His ways are oft past finding out, not because He does not act by law, but because His use of law is then superhuman. And thus it is in the ministry of angels. 'Are they not all ministering spirits sent forth to minister to the heirs of salvation?' (Heb. i. 14.) What a cold and dreary world would this be to the servant of God, if it were subject solely to the blind action of physical laws! a marvellous exhibition of mechanism; but a world without life or generous impulses, without invention, mind, or genius. How does the heart expand to the Scripture view of the spiritual agencies at work, pervading the

* *Reign of Law*, p. 31.

universe, using all the laws with which it is endowed to carry out the Divine commands! 'The chariots of God are twenty thousand, even thousands of angels' (Ps. lxviii. 17). 'His angels excel in strength, they do his commandments, hearkening unto the voice of his word' (Ps. ciii. 20). 'They encamp round about them that fear him' (Ps. xxxiv. 7). 'He giveth his angels charge over them, to keep them in all their ways' (Ps. xci. 11).*

(3.) Lastly: as Scripture never speaks of the order of nature as self-maintaining or absolutely immutable, but as subsisting and upheld by the Divine power (Col. i. 17; Heb. i. 3), and also as subject to change from its ordinary course in miracles when it seems good to the Divine will; † so Science, on the other

---

* Mr. Lecky, in the first volume of his *History of Rationalism in Europe*, in exposing certain extravagant views on this subject held in the middle ages, seems to me, pendulum-like, to go to the opposite extreme, and not to give proper weight to Scripture evidence. So also, in justly calling in question modern miracles, he uses the language of scepticism, though probably not expressive of his own mind, and seems to overlook the evidences for the miracles recorded in the Holy Scriptures. He gives, or at any rate alludes to, the loose reasoning of infidels on the subject, and does not, as he might have done in few words, give the antidote.

† How beautiful is the language of that great Christian philosopher, Dr. Chalmers, as he expatiates upon the constancy of nature's laws. The Constancy of Nature demonstrates the Faithfulness of God, and not that the world is independent of Him:

'In every human understanding He hath planted a universal instinct, by which all are led to believe that Nature will persevere in her wonted courses, and that each succession of cause and effect, which has been observed by us in the time that is past, will, while the world exists, be kept up invariably, and recur in the very same order through the time that is to come. This constancy, then, is as good as a promise that He has made unto all men; and all that is around us, on earth or in heaven, proves how inflexibly the promise is adhered to. The chemist in his laboratory, as he questions Nature, may be almost said to put her to the torture; when tried in his hottest furnace, or probed by his

hand, demonstrates nothing contrary to this, although this school of theorists assumes that it does. The presumption that Nature is uniform in its course is based, not upon any abstract ideas regarding natural laws, but solely on experience; and that experience is grounded upon the combined testimony of observers and experimenters. All, therefore, that Science teaches us is, that within the wide range of modern investigation no departure from uniformity has been detected. No doubt a correspondingly strong argument flows from this against the probability of any departure ever taking place; *unless* it appear that other facts, equally supported by testimony, can be produced which show the contrary. These facts do exist, and the Scriptures,

searching analysis to her innermost arcana, she, by a spark or an explosion, or an effervescence, or an evolving substance, makes her distinct replies to his investigations. And he repeats her answer to all his fellows in philosophy, and they meet in academic state and judgment to reiterate the question, and in every quarter of the globe her answer is the same: so that, let the experiment, though a thousand times repeated, only be alike in all its circumstances, the result which cometh forth is as rigidly alike, without deficiency, and without deviation. We know how possible it is for these worshippers at the footstool of science to make a divinity of matter; and that every new discovery of her secrets should only rivet them more devotedly to her throne. But there is a God who liveth and sitteth there, and these unvarying responses of nature are all prompted by Himself, and are but the utterances of His immutability. They are the replies of a God who never changes, and who hath adapted the whole materialism of creation to the constitution of every mind that He hath sent forth upon it. And to meet the expectation which He Himself hath given of nature's constancy, is He at each successive instant of time vigilant and ready, in every part of His vast dominions, to hold out to the eye of all observers the perpetual and unfailing demonstration of it. The certainties of nature and of science are, in fact, the vocables by which God announces His truth to the world; and when told how impossible it is that nature can fluctuate, we are only told how impossible it is that the God of nature can deceive us.'— *Chalmers' Works*, vol. vii. p. 212.

and the great body of the Christian evidences which encompass the sacred volume, testify to them. The theory of evidence is as much a branch of science as the theories of physics; and it is a matter of demonstration, that the kind of testimony on which the Scripture statements rest, is as sure a ground of belief as that on which the discovery of the uniformity of nature rests. Nothing, therefore, which physical science has taught us, regarding the action of the laws of nature within the range of its observation, can disprove what Scripture testifies regarding a period which did not come within that range.

Thus Scripture and Science lead us to the same conclusion in these points: they both testify to the general uniformity of nature; they both teach us that the world is not abandoned to the unbroken effect of natural laws, but that the interposition of moral agency repeatedly causes the effect of one law to cease, and that of another to commence abruptly; and they agree also in this, that whereas Scripture represents the course of nature to be perpetually dependent upon the Divine power, and as subject to change in miracles, whenever it seems good to the Divine will, physical Science demonstrates nothing contrary to these: in short, the science of evidence establishes the latter.

Difficult as some sceptically-inclined minds find it to believe in any change in the uniform course of nature, there is nothing in itself incredible in the thought, which Scripture presents to us, that the Almighty should show His power, and thus accredit His messengers, by occasionally, as He sees needful, using in a superhuman

Miracles not incredible.

way one or more of the laws which He has impressed on the material world. The possibility of this can be denied only by denying the Divine Omnipotence. It is the assumed absence of this Almighty interposition, and the immutability of the course of nature thence inferred, which leads some men virtually to shut out God from free action in His own world. This is undoubtedly the tendency of the remarks of both the authors I have quoted eleven pages back. One of them, in speaking of the 'grand foundation conception of universal law,' asserts 'the impossibility ... of any modification whatsoever in the existing conditions of material agents, unless through the invariable operation of a series of eternally impressed consequences, following in some necessary chain of orderly connexion, however imperfectly known to us;' shutting out the interposition of the Almighty altogether. And the other states, that 'the student of science is learning to look upon fixed laws as universal,' and that, as a consequence of this, 'miracles are felt by very many to be difficulties in their way.'

I would, moreover, observe, that both of these writers carry our thoughts outwards to a region beyond the limits of human observation. One speaks of 'eternally impressed consequences;' leading us back into eternity past, as a source from which present phenomena take their rise. The other refers to 'higher laws,' beyond the region of human observation; and so carries us into the mysterious labyrinths of the unknown. Now, when we speak of a miracle we do not mean an event which has not been preordained in the scheme of the Divine government of the world, but an event which is clearly contrary to the ordinary course of nature as observed by us. If we could take within the compass of our thoughts the whole order of things, material and spiritual, in heaven and in earth, and all their successive changes in past and future time, including what we call miraculous as well as ordinary events, and were to look upon this as the ordained law of the universe, no doubt no departure from law in this sense could ever take place. For this is the complete expression of the will of Him who rules all things after His own pleasure, and to whom past, present, and future, are all one. But this is an idea of law which no finite being can ever grasp; it is altogether an unpractical idea, which can never serve as a guide to the student of science, or to any other mortal. Law, in this comprehensive view of it, is altogether beyond our knowledge and observation. It is only those few and partial laws, parts only of the universal order of things, which we are practically concerned with. 'These are but parts of his ways,' only 'a little portion of him:' 'but the thunder of his power who can understand?' (Job, xxvi. 14.) It is in these partial laws that that uniformity is discerned, which is borne testimony to by both Scripture and Science. And it is the knowledge of this general uniformity which puts it in our power to pronounce, whether any event is miraculous or not; for a miracle is such, only because it is a departure from this ascertained constancy of the course of nature. It is, therefore, a wrong conclusion to draw from this general uniformity, that under no circumstances can an extraordinary departure from it occur. This would be simply reasoning in a circle. The occurrence of such a departure is a matter of testimony, and not of speculation. It is, then, a begging of the whole question, to say that miracles are impossible, and, in the face of evidence the other way, to base the assertion on the uniformity observed in that part of nature which we have examined. A well-attested miracle can be no real 'difficulty in the way' of the student of science, unless he banishes from his mind all idea of the presence and power of the Almighty in the world. The miracle is, in fact, a special manifestation of God's power and presence; and to attribute it to the 'natural action of still higher laws,' is only an unsuccessful attempt at escaping from that conclusion. For, in the whole range of scientific investigation, no law has ever been discovered, nor anything analogous to it, which would of itself abruptly interfere with the action of another law, in the way that miracles interrupt for a time the ordinary course of nature. All analogy teaches us that such could take place only by the intervention of moral or spiritual agency—the very thing I am here contending for. The difficulty of miracles, as opposed to the belief in the uniformity of nature, lies only in the contracted view which those take who would fain exclude the Almighty from His own world, and deify the laws by which He has been pleased ordinarily to act.

I cannot close these remarks without observing, that, with all our progress in scientific investigation and discovery, the real advance which we have made is but a single step compared with the boundless expanse which lies before us. The time will never come for us to speak

Prospects of Science sometimes over-estimated.

thus proudly of our position. 'It may be indeed,' says Dr. Temple, 'that the scientific student is every day less and less driven to confession of the narrowness of his knowledge; he has less occasion for the humility which once allowed vast realms of nature to lie out of the domain of science, and was wont to say, when baffled, "Here human powers can go no further, this knowledge God has reserved for Himself." On the contrary, he is now inclined to think that, if only time enough be given, there seems to be no kind of phenomenon under the sun which patient study will not bring within the range of science.'* In reply to this I would say, that it is notoriously the experience of the profoundest philosophers, that the further their investigations are carried the more do unsolved phenomena rise up before them: the deeper their research into the mysteries of nature and the wider the extent of their discoveries, the more do they realize the utter hopelessness of their ever finding any limit to the field of their investigations. The more a philosopher knows, the more is he satisfied that the field is infinite in extent; and not only do new facts rise up in every direction, but new wonders. His investigations, limited as they are, and ever must be, bring this advantage; in what he has explored he traces the footsteps of the Supreme Intelligence. He sees them in the order, harmony, adaptation, and beauty, which characterize what he does know of His works; but he is no nearer to the First Great Cause Himself. He learns a lesson, as he advances, of increasing humility—he knows more of the majesty and infinity of the Creator the further he

* *Sermon before the British Association.*

penetrates; but is convinced that, whatever wonders may be unfolded to his admiring gaze, regions illimitable lie beyond. It is not that he feels his impotency in having no power to 'alter one law,' 'to create or to annihilate' one atom of matter; he is far from having reached such an eminence as to have no other calls to humility but these. He is still surrounded by phenomena innumerable, whatever progress he has made, which indicate laws within laws, and laws beyond them in endless succession. And to conceive that he will ever attain to such a knowledge of law as to comprehend all the courses of nature; that he will, as it were, be able to predict the ways of God by having unravelled the mysteries of those inexorable laws by which He is, as this theory of law seems to assert, inevitably bound—is a thought the possession of which would deprive him of every claim to the title of philosopher.

There are but few laws in the physical world which have been discovered and demonstrated. Phenomena innumerable have been observed and recorded and arranged in classes; many, at first thought independent, have been grouped together as evidently flowing from the same causes; but those causes are rarely known or understood. Laws, or rather properties of matter under various circumstances, have been partially and empirically made out; but few laws, properly so called, have been demonstrated. Of these the most comprehensive and conclusively established is the Law of Universal Gravitation. But as to the physical cause of this law, we are utterly at a loss to suggest even a conjecture! What can be the connexion between an atom in the sun and an atom of this earth? and yet, according to the Law of Universal Gravitation, they attract each other even through all intervening matter, and that with precisely the same force as if that matter did not intervene. Innumerable phenomena in light, and of great variety and beauty, can be explained upon the hypothesis that light is caused by the vibratory impulse upon the retina of an elastic medium pervading space. The Undulatory Theory of Light may, therefore, be considered to be one of those few laws which have been proved to be laws of nature. But there are many questions unsolved regarding the constitution and properties of this medium. And so is it in every department of science. The few laws we have determined are comprehensive facts, discovered empirically and considered to be true, only because of the multitude of cases they will group without any material opposing circumstance. They are our fixed points, upon which we can stand in security, while, however, we are surrounded by a restless ocean of unsettled questions and undetermined phenomena to which there is no limit, and never will be.

Dean Mansel has advanced an argument against these theorists which is so effective, that I cannot abstain from transferring it to my pages. In reply to the boast that 'the inevitable progress of research must, within a longer or shorter period, unravel all that seems most marvellous,' he remarks, in reference to miracles, 'that the fact of a work being done by human agency places it, as regards the future progress of science, in a totally different class from mere physical phenomena. The

Progress of Science confirmatory of Miracles.

appearance of a comet, or the fall of an aërolite, may be reduced by the advance of science from a supposed supernatural to a natural occurrence; and this reduction furnishes a reasonable presumption that other phenomena *of a like character* will in time meet with a like explanation. But the reverse is the case with respect to those phenomena which are narrated as having been produced by *personal agency*. In proportion as the science of to-day surpasses that of former generations, so is the improbability that any man could have done in past times, by natural means, works which no skill of the present age is able to imitate.'* The progress of science, therefore, so far from encouraging any hopes of our being hereafter able to account for miracles upon physical principles, only enhances their miraculous character. No more effectual blow could be given to their whole argument. There is an air of assumed superiority in the manner in which these views of the supremacy of law, at the expense of miracles, have been put forth. But we may feel support in the fact that not one of our first-class philosophers has ever advocated these views, and that the prince of philosophers—who, from the vastness of his intellect and the power of penetration into the secrets of nature which he possessed, would be under a stronger temptation than any other man to set up this hope in prospect, if there were any real foundation for it—has given his decisive verdict the other way, and has described the Deity as 'secundum leges accuratas con-

---

* *Aids to Faith*, p. 14. See also the Duke of Argyll's *Reign of Law*, chap. i., on the difference between the 'supernatural' and 'superhuman.'

stanter co-operans, *nisi ubi aliter agere bonum est;*' that is, as acting according to exact laws, *except when it be good to act otherwise.*

13. Another illustration of my argument may be drawn from the objections which have of late years been brought against the manifestations of Design in the works of creation, on which the Doctrine of Final Causes is based. Scripture teaches us plainly that God has had a purpose in all His works. This is one of the lessons taught us in the brief space of the First Chapter of Genesis. We see it in the very framework of the chapter. Every day's work is pronounced to be 'good,' and that of the sixth to be 'very good' (i. 31); which evidently means good for the purpose for which each day's work was performed. And more than this, it is said, 'God made the beast of the earth *after his kind*, and cattle *after their kind*, and every creeping thing that creepeth upon the earth *after his kind*' (Gen. i. 25); and so of the grass, and the herb, and the tree. Design is stamped upon the whole transaction. In one or two instances also it is distinctly stated what the end in view was. The two great lights, or light-holders, were made in order 'to divide the day from the night;' 'for signs, and for seasons, and for days, and years;' 'to give light upon the earth;' 'the greater light to rule the day, and the lesser light to rule the night' (i. 14–16). He made also herbs and trees for the use of man, and other living creatures (i. 29, 30). And in the next chapter we are told why woman was created; viz. to be a 'help-meet' and companion of man (i. 18–24). So that from the very beginning Design is prominent, when God describes His own works. 'He created not

the earth in vain; he formed it to be inhabited' (Isa. xlv. 18).

The whole of Mr. Darwin's theory of natural selections, which I have already considered, is opposed to this view. He regards all plants and animals as having been progressively developed by accidental changes from previous forms, and all originally from one form; and this theory he holds without one example, either in the existing or the fossil world, of a plant or an animal having descended from any but progenitors of its own species.* Whereas all sound naturalists have held, that 'the power and wisdom of the Creator designed and made organic beings for certain objects; and in the inimitable wisdom, skill, and beauty of the contrivances by which this has been effected, we see indisputable proofs of the high source of the general design.' And in the immutability of species 'we see so clearly an additional proof of the origin of the design, that we are disposed to look on this law as the very sceptre of the Creator—as visible evidence that the Supreme Intelligence which invented all the varieties of life has resolved that the original plan should be maintained in all its purity, and that the boundary lines of separation should be perpetually respected.'†

* Even Professor Huxley, Mr. Darwin's strenuous advocate, is obliged to confess this. 'After much consideration,' he writes, 'and with assuredly no bias against Mr. Darwin's views, it is our clear conviction that, as the evidence stands, it is not absolutely proven that a group of animals, having all the characters exhibited by species in nature, has ever been originated by selection, whether artificial or natural.'—*Lay Sermons, Addresses, and Reviews*, 1870, p. 323.

† See *The Darwinian Theory of the Transmutation of Species Examined*, by a Graduate of the University of Cambridge, p. 37. An excellent work, the writer of which is evidently well qualified for the

The views set forth by Mr. Darwin are the same as those advocated by M. Géoffroy Saint-Hilaire in opposition to Cuvier, though illustrated by a far larger amount of facts than had then been accumulated; put together, however, with so many conjectures and assumptions as to make the theory no more trustworthy than it was in Cuvier's time. Cuvier enunciated his leading principle in the following terms:—' Zoology has a principle of reasoning which is peculiar to it, and which it employs with advantage on many occasions: this is the principle of *the conditions of existence*, vulgarly called the principle of *final causes*. As nothing can exist if it do not combine all the conditions which render its existence possible, the different parts of each being must be co-ordinated in such a manner as to render the total being possible, not only in itself, but in its relations to those which surround it; and the analysis of these conditions often leads to general laws, as clearly demonstrated as those which result from calculation or from experience.'* Dr. Whewell shows that this principle, self-evident as it appears to some, has been in the hands of Cuvier an instrument of important discoveries. In the Introduction to his great work on *Fossil Remains* he says, 'Every organized being forms an entire system of its own, all the parts of which mutually correspond, and concur to produce a certain definite purpose by reciprocal reaction, or by combining to the same end. Hence none of these separate parts

task. With a masterly common sense he detects the contradictions, inconsistencies, and folly of the system.

* Quoted by Dr. Whewell, *History of the Inductive Sciences*, vol. iii. p. 472.

can change their forms, without a corresponding change in the other parts of the same animal; and consequently each of these parts, taken separately, indicates all the other parts to which it has belonged. Thus, if the viscera of an animal are so organized as only to be fitted for the digestion of recent flesh, it is also requisite that the jaws should be so constructed as to fit them for devouring prey; the claws must be constructed for seizing and tearing it in pieces; the teeth for cutting and dividing its flesh; the entire system of the limbs or organs of motion for pursuing and overtaking it; and the organs of sense for discovering it at a distance. Nature must also have endowed the brain of the animal with instincts sufficient for concealing itself, and for laying plans for catching its necessary victims.' By such considerations Cuvier was able to reconstruct the whole of many animals of which parts only were given; 'a positive result,' Dr. Whewell adds, 'which shows both the reality and the value of the truth on which he wrought.' This principle enabled him to classify animals in a way never before effected; and his immense knowledge and powers of thought led him to results, the value of which every naturalist knows, under its master-guidance. The opposite hypothesis, that living things have attained their present state by a blind natural-selection without forethought or design marking the hand of a Supreme Intelligence, is a dream of the imagination, and will not stand scrutiny. The school which advocates it, does indeed admit that Order and Beauty prevail in what *we* call created things, although they attribute this to chance.

There is, however, another school which seems even

to deny this order and beauty; and to say that there is nothing whatever in the world and the arrangement of its parts to indicate design. Compte and his followers, the school of Positivists, assert this. They presumptuously repudiate faith and revelation; and profess to build only on absolute knowledge, or science, and nothing else. According, then, to the Positivists, as well as the followers of Mr. Darwin, Scripture and Science are at variance in this idea of Design.

But it is not difficult to see where the fallacy in Compte's reasoning lies. He cannot but acknowledge that there are innumerable instances of fitness in the physical world of part to part, and that also even within our own frames. The uses of these adaptations are beyond question. But there are countless instances in which we cannot discern the final cause or object. There are those who lean to their own conceits, and think they could have framed the world in some respects better than God has made it. My readers will remember that Buffon ridiculed the idea of the sloth being the work of a Supreme Intelligence, and called it a mistake of nature. But comparative anatomy has since shown how wonderfully adapted its structure is to its habits. Laplace laughed at the idea of the moon having been made to give light upon the earth by night, observing that if that had been the purpose it would have been placed at such a distance as always to be opposite the sun, and therefore appear full all night and every night. But he omitted to state, that that distance would be so much greater than at present as very considerably to diminish the amount of light we receive from the moon, even to 1-30th of its present quantity; nor can we tell the benefits the earth derives from a periodical cessation from light of both sun and moon. An article not long ago appeared entitled 'The Waste of Space,' with an attempt to build up an argument against design; substituting ignorance for knowledge; and really inferring that because we do not know the use of space in all its expanse, therefore it is useless and is wasted. So the mistake of Compte was this most illogical step, viz., placing in his argument the known and the unknown on the same footing. If ten thousand cases of adaptation had been discerned, and ten thousand other cases in which no adaptation has yet been discovered, these would, in his mode of reasoning, cancel the former and leave a perfect blank! No doubt we are often baffled in attempting to discern the uses of some things in creation. In consequence of this want of sagacity to trace the final cause in every instance, it has been proposed to base the illustration of the workmanship of the Divine Hand in the unity and beauty of plan, the symmetry and order which we see prevailing in the works of creation. Both indeed may be used, as far as each will carry us; and the positive knowledge they give is a certainty which no amount of ignorance can cancel. And judging of these works as we should judge of any human works, we cannot, unless we wilfully blind our eyes, fail to see the indications of supreme skill and intelligence. Thus all God's works praise Him. Illustrations of this from Galen down to Paley and many modern investigators are countless.*

---

* See some striking examples in Mr. Darwin's book on the *Fertilisation of Orchids*, where design is seen in endless ways, though he attributes it all to development. See also some equally striking

The manifest proofs of design which the world exhibits show that the God of the Scriptures is the Creator of the world and of all that is in it. Wherever we trace the manifestations of design, there we have the indications of His workmanship. This is positive knowledge. But if we fail to trace His hand in other works and infer that He is not there, we betray our forgetfulness that we are but finite creatures, and are not to measure the divine works by our ignorance; however much, through His goodness, we may be able to unravel by the successful use of the powers He has given us.

But beyond this palpable fallacy in the reasoning of Compte—a fallacy of which he would have been ashamed in the ordinary affairs of life—an equal fallacy lies in the theory of the universe which he adopted. He knew too much of the wonderful structure of the world and all it contains to believe that this beautiful system is the product of chance. He knew too much of its real adaptations to adopt the Epicurean doctrine. He attributes all that we see to the action of law—of properties inherent in matter. And so do we. But there he stops. We advance further. With marked inconsistency he ventures thus far beyond his prescribed limit of what he considers to be absolute knowledge, and ventures upon an hypothesis, but will not advance. We go a step further, and ask, Whence these properties? And reason is gratified with the thought, that here are the footsteps of a Supreme Intelligence. Here is much that is inexplicable, though not neces-

examples in the Duke of Argyll's *Reign of Law*. There is a remarkable illustration of design in the structure of the kangaroo and its young, given by Professor Owen in *Phil. Trans.* 1834.

sarily contradictory. But there is a vast field of positive knowledge, which at every step confirms our conviction that an Almighty Hand has framed and fitted the universe, and seals the harmony between Scripture and Science.

The assumption that these properties have not been given to matter, but are really inherent—and, in fact, are eternal — is simply gratuitous. These so-called philosophers are like grown-up children still in the elements of knowledge, and seem, through their wilful blindness, unable to take the next and higher step in learning, which the facts around them teach, as well as the Word of God.

It might be shown that their dependence on physical fact alone is very hollow ; for mental and spiritual knowledge, supported by suitable evidence, is as true and substantial as any physical knowledge of the external world. If this is ignored, and the evidences of revelation are despised, of course men will lose themselves in the labyrinths of their own folly and presumption ; they 'darken counsel by words without knowledge,' and can be awakened from their insensibility only by a visitation from on high, as 'the Lord answered,' confounded, and humbled Job and his companions 'out of the whirlwind.' (Job, xxxviii. 1, 2.)

Professor Huxley's volume on *Man's Place in Nature* is a treasury of facts which strikingly illustrate this part of my subject, as I have already shown regarding his protoplasm—viz., that Design is to be traced throughout creation, though this is not the use he makes of his own investigations. In a most lucid manner he explains the results of recent physiological research; and tells us that most animals, and it may in the end be shown that all animals, have their beginning in an *ovum*, or egg.* That this egg, except the bird's, is a minute object; in the cases of dogs and men (the examples he specifies), from $\frac{1}{120}$th to $\frac{1}{130}$th of an inch in diameter. In its development it is long, he tells us,† before there is any perceptible distinction between the embryo of the dog and of the human being; and at the stage at which a difference is discernible between them, there is none between those of the ape and man. 'Exactly in those respects in which the developing man differs from the dog, he resembles the ape. . . . It is only quite in the later stages of development that the young human being presents marked differences from the young ape, while the latter departs as much from the dog in its development, as man does.' He then carries us through an interesting comparison between man and some different kinds of apes in their adult state, comparing them in their spinal column, their back-bone, ribs, pelvis, skull, teeth, extremities, hand and foot, brain. The result is, 'that the structural differences which separate Man from the Gorilla and the Chimpanzee are not so great as those which separate the Gorilla from the lower apes.'‡ Now, there are two conclusions to one of which this investigation, thus far carried, might lead us, had we no other guide: viz., either that animals may all have had a common origin, or that the Creator has been pleased to follow a plan or type in bringing creatures into existence. Which does Professor Huxley take? He tells

* *Man's Place*, 3rd ed. p. 60.   † Ibid. p. 66.
‡ Ibid. p. 102.

us of 'the great gulf which intervenes between the lowest man and the highest ape in intellectual power:'* of 'the immeasurable and practically infinite divergence of the human from the simian stirps.'† Again he says, 'No one is more strongly convinced than I am of the vastness of the gulf between civilized man and the brutes; or is more certain that whether *from* them or not, he is assuredly not *of* them. No one is less disposed to think lightly of the present dignity, or despairingly of the future hopes, of the only consciously-intelligent denizen of the world.'‡ Which side, then, of the alternative does he choose? That which is really the most illogical, and which also is in direct contradiction to Scripture. Why the most illogical? His conclusion, after speaking of the *ovum*, is this: 'Startling as the last assertion may be,' (that which I have quoted regarding the affinity of man to the lower animals) 'it is demonstrably true;' and he then adds, 'and it alone appears to me sufficient to place beyond all doubt the structural unity of man with the rest of the animal world, and more particularly and closely with the apes.'§ And that we may not mistake his meaning, I will quote further: 'If man be separated by no greater structural barrier from the brutes than they are from one another, then it seems to follow that if any process of physical causation can be discovered by which the genera and families of ordinary animals have been produced (alluding to Mr. Darwin's theory), that process of causation is amply sufficient to account for the Origin of Man.'‖ And what does this rest upon? The

---

* *Man's Place*, p. 103. † Already quoted at p. 239, Note.
‡ *Man's Place*, p. 110. § Ibid. p. 67. ‖ Ibid. p. 105.

failure of his powers, and of any man's powers, to discern any difference in the earliest stages of development of the dog, the ape, and man, especially of the last two. But does not the issue show him that there *is* a difference, although its existence at first is not discernible to our feeble faculties? The *ovum* of the dog becomes a dog; that of an ape, an ape; and of man, a man. The *ovum* of a dog never becomes a man; nor that of a man an ape or a dog. This, his own chosen illustration of development, should surely have convinced him of the fallacy of any such theories. Apparently alike as individuals of different races are in their beginnings, so as not to be distinguishable to man's perception, yet even those individuals develop into things perfectly distinct, each after his kind. Though the beginnings *seem* to be the same, the individuals in their development never change from one to the other; and each preserves its own line, always unbroken. It is marvellous how logic seems to fail men, often of great research! In the presence of this alternative—Design or Common Origin—blinded by the obscurity which hangs over the earliest stages of life and growth, the speculator forsakes his guide, the reason with which he is endowed, takes a prodigious leap in the dark, and finds himself landed in the theory he so desires to believe true; whereas the more sober and steady philosopher views the whole from beginning to end; uses the facts so industriously brought together; recognizes the similarity which runs through the earliest growth and final structure of the animal world; sees, however, on the one hand, the 'immeasurable and practically infinite' distance between man and the nearest of other creatures to him; but, on the other, marks that in mere structure he is nearer to the nearest than they are to other animals; and draws this conclusion, that Design is eminently stamped upon creation, and also that as man and the nearest animals to him cannot possibly have had a common origin, because of the 'immeasurable and practically infinite' distance between them, so, as to other races of animals, no satisfactory evidence can be drawn from similarity of structure that they must have had a common origin, as it has been proved that this cannot have been the case with creatures nearer even than they are in this respect, namely, man and the animals which most resemble him, but that they are made upon a common plan, which circumstance is one of the surest indications of Design. He rejoices in this conclusion, for it agrees with Scripture, and one more evidence is supplied that impartial investigation will always show that Scripture and Science are not at variance.

Before concluding this section I would observe, that there is one effect of these views, which exclude or enfeeble the perception of Design in God's works and virtually deny that God is the Author of Creation, against which all need to guard themselves; as it has grown very much into a habit, even among those who would shrink from avowing such views as their own. I mean the habit of personifying Nature, of in fact deifying Nature, of attributing to Nature, and to nothing beyond Nature, the marvellous phenomena which science brings to light. A pantheistic halo of glory is thrown around the works of God, and God is forgotten or never mentioned. This reminds

Use of the Term Nature.

us of those who 'became vain in their imaginations, and worshipped and served the creature rather than the Creator' (Rom. i. 21, 25). It may be attributed in some cases to an ill-placed notion of reverence, shrinking from a too familiar mention of the name of the Most High. Never should we take that Name upon our lips in vain. But this is no excuse for that almost habitual exclusion, which you see in some writers and speakers, of God Himself from the world He made and perpetually upholds. How different is the temper of mind which delights to acknowledge God in all His works, and to remember, that 'He it is who maketh Arcturus, Orion, and Pleiades, and the chambers of the south, who doeth great things past finding out; yea, and wonders without number. . . . HE maketh small the drops of water; they pour down rain, . . which the clouds do drop and distil upon man abundantly. . . . THE LORD hath His way in the whirlwind and in the storm, and the clouds are the dust of His feet. Thou makest the outgoing of the morning and evening to rejoice. . . Thou visitest the earth and waterest it. . . . Thou preparest them corn. . . Thou crownest the year with thy goodness, and thy paths drop fatness. . . . O Lord, how manifold are Thy works! in wisdom hast Thou made them all: the earth is full of Thy riches.' (Job, ix. 9-10; xxxvi., 26-28; Nah. i. 3; Ps. lxv. 8-11; civ. 24.)

14. The last example which I shall adduce is in some respects more remarkable than any of those which have preceded; not, however, from the formidable nature of the difficulties advanced. It may be said that Dr. Colenso's

<small>Arithmetical objections to the Pentateuch.</small>

arguments against the truth of the Pentateuch are more philological than scientific, and therefore hardly come within the scope of this treatise. But as the argument of the first part turns almost entirely upon arithmetical comparisons, and as his objections have been widely made known, I willingly add the illustration which they so abundantly afford of my leading principle.

(1.) In the enumeration of the seventy persons of Jacob's family who went down into Egypt when Joseph was governor, two grandsons of Judah are mentioned, who, Dr. Colenso maintains, could not have been born at that time. Various hypotheses have been framed to remove the difficulty, but none of them satisfy him, and he accordingly pronounces the Pentateuch to be a work devoid of historical credit!

In reply, I would say, that the Pentateuch is not received by us as true upon evidence of this kind. It is too ancient a document to stand or fall by such tests. There is a wide difference between obscurities and contradictions. Experience teaches us how in some instances the discovery of an apparently trivial fact will altogether change the complexion of a narrative. The alteration, too, through copyists or otherwise, of a single word, may, in places affected by it, irremediably confuse the chronology without invalidating the history in which it occurs. By the application, then, of such tests, we may, perhaps, detect where obscurities lie, or where error has crept in; but if we judge the document itself, ancient as it is, by this arithmetical process, we may be led to an utterly false conclusion, and shall be rejecting as untrustworthy what is in reality truth. If it were asked, What is the difference between this admission, that the effect of miscopying and of present ignorance may, in some few instances, affect the reliability of a statement in Scripture, and the notion that the writers were not so entirely inspired as to preserve them from error of every kind, I should reply, that the difference is wide indeed. No doubt, that in the chain of communication between the Infinite God and ourselves, the human element must come in somewhere; but it cannot require much thought to perceive, that error in the mind of the original writers would lead to a far more serious degree of uncertainty in their writings, than the natural fallibility of careful copyists could introduce. Error in the one case might be worked everywhere into the texture of the writing, and lie unperceived; whilst mistakes in the other case, by their very abruptness, awaken suspicion and lead to their detection in the few places where they occur.

Dr. Colenso's first objection illustrates these remarks. It presents, no doubt, a real difficulty, one which has long been observed. In examining a document in which events and dates incidentally mentioned hang together, and by various undesigned coincidences bear testimony to its general truthfulness, but in which there is somewhere an omission of some fact, or a miscopying of some word, a mathematician might well employ himself in determining where, according to the theory of probabilities, the source of difficulty most likely lies. But whether he fail or succeed, to argue from the existence of such a difficulty, that the original narrative is not historically true, is a summary way of disposing of the matter, which will not satisfy a philosophical mind.* For testing the truthfulness of the original document and the whole Pentateuch, I should not be at all anxious to clear up such a difficulty; though to free the received text of the smallest errors, and to elucidate obscure passages by a reference to ancient history, manners, and customs, must always be desirable.

One solution of the difficulty Dr. Colenso first propounds is this: that though Hezron and Hamul, Judah's grandsons, may have been born in Egypt, the fact of the death of Er and Onan, his eldest sons, being recorded in close connexion in the same verse, with the introduction of these names (Gen. xlvi. 12), is a clear intimation that Hezron and Hamul were taken as the representatives of their deceased and childless uncles; and the reason why they were singled out, although other children were no doubt born in Egypt, is that they were heads of families (see Num. xxvi. 21). Dr. Colenso, in transcribing the verse in which Hezron and Hamul are mentioned, has been guilty of a mistake which he of all others should have avoided. He omits the small but important 'were,' which is distinctly expressed in the Hebrew as well as in our English translation; its introduction and the

* Mr. Isaac Taylor says, 'A narrowness of view—a sideway of blindness—right and left, has often been the misfortune of those who have signalized their ability on single lines of mathematical science. Very eminent minds have shown this want of breadth or grasp. Such men may be mathematicians, but philosophers they are not; and so it is that the same quality which constitutes their eminence, while they are on the single path which they tread, shows itself only as pedantry whenever they step off that path. Men of this order are peculiarly unfitted for conducting inquiries of a mixed historic kind.'—*Considerations on the Pentateuch*, p. 22.

consequent change in the mode of bringing these names into the genealogy obviously confirms the interpretation above adopted.

There is no difficulty in understanding why children, though born in Egypt, should, when they were selected to fill up a broken genealogy, be enumerated with those who belonged to Canaan and came down into Egypt, if they were the children of marriages contracted before leaving Canaan, and were born while the exiled colony kept together in its unity during the lifetime of its head. Jacob was seventeen years in Egypt (Gen. xlvii. 28), which would allow sufficient time for the birth of Hezron and Hamul. It is no valid objection to this view, that the sacred writer expressly states, that Joseph's sons were born in Egypt; for that statement seems rather to point to the fact that Joseph their father did not belong, as the other sons of Jacob did, to the colony which migrated into Egypt. Indeed, the writer seems directly to recognize the principle I have advanced; for although he explains that Joseph's children were born in Egypt (xlvi. 20), he yet enumerates them (27) among those who came into Egypt, no doubt because their father came into Egypt, though at an earlier time. The contradiction, if there really is one in the narrative, would be so obvious, that there can be no doubt the language is perfectly accordant with the usage of the Jews in making up their genealogies, though we may not know precisely what their usage was.

(2.) Dr. Colenso's next objection, that it was impossible that 'all the congregation' could be gathered at the 'door of the tabernacle' (Lev. viii. 3, 4) because it was too numerous, may be summarily dismissed, notwithstanding his arithmetical calculations, with a reference to the text, 'Then went out to him Jerusalem and all Judea, and all the region round about Jordan' (Matt. iii. 5), as well as the ordinary language of modern times. Nobody conceives that every individual in those countries flocked to the Baptist, but persons out of all parts of them. So persons representing all the congregation, no doubt, appeared before the tabernacle.

(3.) The third objection, that Moses and Joshua addressed all Israel (Deut. i. 1; v. 1; Josh. viii. 34, 35), whereas, it is impossible that so vast a number could have heard, is as easily disposed of. Could not Moses and Joshua speak to the people through the officers of their tribes? Does the commander of a great army never address his troops, who are too numerous to hear? Is not his address read to them at the head of their companies?

(4.) The next objection is, that the dimensions of the camp were so vast, that the priests could not possibly perform the work assigned them of carrying out the refuse of the sacrifices beyond the camp (Lev. iv. 11, 12), nor the people perform their daily necessities as commanded (Deut. xxiii. 11-14). The reply is, that the Hebrew word 'carry' is in the causative form, and that the priests were to cause to be carried out, that is, to send out, the refuse of the sacrifices. Moreover, there is no reason for supposing that the twelve encampments were not separated by wide spaces or streets, which would be reckoned 'without the camp.' This would at once diminish the distance to between one-third and one-fourth part of the estimate when the whole is supposed to consist of one large square.

(5.) His next difficulty is, that the sum total of the males of the whole congregation was identically the same at two periods six months apart (Ex. xxxviii. 26 ; Num. i. 46); and that as no census is mentioned the first time, but one is on the second occasion, it looks as if the first is borrowed from the second, and that this destroys the historical verity of the whole. In reply it may be said, that though no census is mentioned on the first occasion, it does not follow that none was made. Nor is it wonderful that the numbers should be the same, especially as it is evident that they were counted by fifties, the portions of the last fifty being omitted or made up to fifty. Moreover, if the number on the first occasion is altered by some copyist to conform it to the second, it does not prove the unhistorical character of the original Pentateuch. Again, why cannot we conceive that the roll made on the first occasion was used on the second to call the tribes over at their great gathering ?

(6.) Dr. Colenso's next difficulty about the number of tents the Israelites would require and the beasts to carry them, betrays an ignorance of Oriental habits. Ten or more persons will sleep under a tent made of a single cloth, supported by a horizontal and two pairs of cross sticks, which one of their number could carry on their march.

(7.) He next asks, how so great a multitude could procure their arms ? (Ex. xiii. 18.) Why not from the Egyptians, by whom they were favoured in their escape ? (xii. 36.) Nor does the text imply that every man was armed, but only a sufficient number to give protection. Moreover, the word is of doubtful meaning, and does not necessarily mean 'armed.' In our version, it is 'harnessed,' and in the margin 'five in a rank,' and may imply merely that they went out in some orderly manner.

(8.) How, he asks, could such a multitude keep the Passover and make all their preparations for departure in a 'single day?' How borrow as they did on the spur of the moment? How gather in one single spot, Rameses (Ex. xii. 37), and march out at such short notice? Whence procure lambs sufficient for the Passover? There is no proof that they did march out at a day's notice. Four days' warning had they at least, and perhaps several more. Moses may have received the command regarding the keeping of the Passover nine days before the tenth day of the month (see Ex. xii. 3, 6); so that, though their haste was great, such that they had not 'prepared for themselves any victual' (xii. 39), it was not so hasty as Dr. Colenso states. And if it had been, the pressure and help the Egyptians gave, combined with the extraordinary circumstances, are sufficient to account for the event. Moreover, they did not necessarily all assemble at one spot; Rameses, in Exod. xii. 37, means the *country*, not the city, Rameses.*

Dr. Colenso bases his assertion that the Passover

---

* 'The word Rameses is used to denote both *a city* and *a country*. In the Hebrew Bible the distinction is clearly marked by a difference in the points.' In Gen. xlvii. 11, where the 'land of Rameses' is expressly mentioned, and here in Ex. xii. 37, it is pointed in one way; and in Ex. i. 11, where the city is spoken of, it is pointed otherwise, and is, indeed, translated differently in our English version, viz. Raamses, not Rameses.— See Bishop Ollivant's (*First*) *Letter on Dr. Colenso's Part I.* p. 49.

was kept the very night after the Almighty instructed Moses regarding its celebration, upon these words, 'I will pass through the land of Egypt this night' (Exod. xii. 12). But an unprejudiced reader of the whole passage from verse 2 to verse 12 inclusive, will not fail to observe that by 'this night' is meant the night which was then being described, that is, the night after the 14th day of the month, and not the night after the words were uttered, which was before the 10th day ; it is the same, in fact, as 'that night' in verse 8, the change of language from 'that night' to 'this night' not being an unusual mode of marking emphasis. Nor does the 'midnight' in which Jehovah declared that the first-born should die in the land of Egypt (xi. 4) necessarily mean the midnight immediately after the words were uttered.

(9.) His next difficulty is, how the Israelites could support their sheep and cattle in the desert. He assumes that as no miraculous means are mentioned, like the manna for the people, none were used. This is altogether gratuitous. Moses does not mention every incident or transaction which occurred. Indeed, all through Dr. Colenso's book a reluctance to admit of the supernatural character of the Exodus and its attendant circumstances betrays itself, and that great event is treated as merely an ordinary transaction with no extraordinary helps. Were we, however, driven to natural causes alone for an explanation, I do not conceive that any charge of impossibility would be established against the narrative. Dean Stanley states that at this day the valleys in this 'vast and terrible wilderness' are still verdant in some spots, and that 'there is no doubt that the vegetation of the valleys has considerably decreased.' Moreover, why could not the bulk of the cattle be pastured in the fertile lands bordering on the desert?

(10.) He next wonders how so vast a multitude could leave room for any part of the land to become desolate, and for the wild beasts of the field to multiply (Exod. xxiii. 29). But Dr. Colenso's mensuration refers to Palestine only. The limits, however, of the promised land were 'from the desert to the river' (verse 31), where by 'the river' is meant the Euphrates, as we learn from Gen. xv. 18; also from Deut. i. 7; xi. 24. And this extent of territory was embraced in the kingdom of David (2 Sam, viii.) and of Solomon (2 Chron. ix. 26, margin).

(11.) Another objection is based upon a comparison of the first-born males with the number of males above twenty years old, which he asserts gives an average of forty-two sons to each mother. After the deliverance from Egypt the first-born were to be dedicated to the Lord (Exod. xiii. 1, 2). Subsequently to this the tribe of Levi was separated for the service of the tabernacle, and was taken as a substitute for the first-born (Num. iii. 5–13). On this occasion the first-born and the Levites were numbered, and found to be 22,273 and 22,000; the first-born exceeded the Levites by 273, and these were redeemed with money, at the rate of five shekels a head (verses 39, 43, 46, 47). The males of twenty years old and upwards being taken at 600,000, there may have been 900,000 males in all; of these, 22,273 being first-born, if we divide the first by the second, we have 40 (the Bishop says 43) for the average number of sons in each family, or, as he says, to each mother—a result so extravagant that Dr. Colenso assumes that the Pentateuch is not true. But what first-born were numbered? Not all that were then living among the 900,000 males, but only those who had been born since the deliverance from Egypt, and were therefore young children, as from that date the dedication of the first-born began. This at once removes the difficulty. An incidental confirmation of this is seen in the price put upon the first-born who were to be redeemed with money, viz. five shekels apiece: for that was the price put upon young children of five years and under, when they were vowed, or sanctified, and set apart, as the first-born were (Exod. xiii. 2, 12), see Lev. xxvii. 6.* The minute internal consistency, and not the untruthfulness, of the various books of the Pentateuch, is most strikingly brought out by this attack upon its veracity.†

(12.) He next objects that the vast multitude of

---

* See Rev. F. W. Fowler's *Answer to Dr. Colenso*, p. 40.

† Should it be thought that the sacred text will hardly admit of the interpretation, that 'only those who had been born since the deliverance from Egypt' were counted, I may refer to the Rev. T. R. Birks' *Exodus of Israel* for another solution of Dr. Colenso's difficulty. He considers, as I do, that *all* the first-born males in Israel were not counted. As the angel of death in the tenth plague of Egypt is supposed to have smitten only the firstborn son in each family, and not the father of the family, even though he may have been the firstborn of his mother, so the same principle is taken to limit the firstborn now counted. Also another limitation arises from the fact that they were not only to be firstborn sons, but the eldest children of their families, having no elder sister (Exod. xiii. 2). A third limitation he derives from the proportion of those who would die in early infancy, and not be counted. In the end he reduces the average number of sons to each mother to between four and five.

600,000 males above twenty years of age, at the time of the Exodus, could not possibly have sprung from Jacob in four generations—the fourth generation from Jacob being that in which they were to come out of Egypt (Gen. xv. 16). During Jacob's lifetime his twelve sons had fifty-three male children, making an average of four and a half each (Gen. xlvi.) This he assumes to be the average number of sons during the generations which follow from that time to the Exodus. But Jacob's sons may have had more children after Jacob's death. His daughters, too, may have had sons, who are here altogether overlooked. Again, although the prophecy was literally fulfilled in the line of Moses and Aaron, that the return to Canaan should be in the fourth generation (Num. xxvi. 59), it by no means follows that in *every* line of Jacob's descendants the generations were only four. Thus in Joshua's line there were ten: see 1 Chron. vii. 22–27. Both these suggestions, regarding the number of children and the number of generations, throw out Dr. Colenso's calculations. Again, he has made no account of the effect of polygamy and concubinage in increasing the population. The same kind of answers may be given to his objections to the number of descendants of Dan and Levi at the time of the Exodus, with the additional one that generally the sons recorded in the genealogies are such as became the heads of families, other sons not being mentioned.

(13.) The number of priests at the celebration of the Passover Dr. Colenso considers to have been quite insufficient for their duties. But we have no means of knowing how many there were. The sons of Aaron are not all necessarily recorded—nor are his daughters;

he may have had many grandsons, for at the Exodus he was eighty-three years of age (Exod. vii. 7). Unnecessary difficulties about the number of birds needed for the offerings are started. He wonders how the multitude of animals whose blood was sprinkled on the altar could have been crowded into the court of the tabernacle; whereas they may have been slaughtered outside the court and brought to the priest in turn before the altar. All these matters are reduced to numbers by Dr. Colenso's arithmetical processes, and from his results he pronounces that without doubt things impossible are told us in the Pentateuch, and that the Pentateuch is therefore unhistorical, untrue.*

But enough of this 'hunting of shadows,' as it has been called. 'By this time' — to use in part Dr. Colenso's own words—'surely great doubt must have arisen in the minds of most readers,' not, as he proceeds to say, 'as to the historical accuracy of sundry portions of the Pentateuch' (p. 90), but as to Dr. Colenso's capacity for examining its historical credibility. If these are the only results he can arrive at, with all the last refinements of German rationalism in his hands, and with a mind strongly biased towards the scepticism of the day, Scripture and Science may rise in triumph under this last attack upon their perfect harmony, and still continue to defy all attempts to set them at variance.

* The reader may consult, with great profit, Mr. Isaac Taylor's *Considerations on the Pentateuch*, especially pp. 31–59, and the *Notes*, third edition, in which he strikingly points out the gift of management, and instinct of order, which have always been the distinction of the Jewish people, and have made it easy to them, and to their official persons, to give effect to religious observances, which to us, at this time, may seem exceedingly difficult.

Although I have endeavoured to give, as so many others have well done, categorical replies to Dr. Colenso's objections to the Pentateuch, so mischievous to the unlearned and sceptically inclined, I must in candour say, that I feel that the objections, as arguments against the truthfulness of the Pentateuch seem to me in many cases so frivolous, and in all so unworthy of a Christian scholar, that they do not deserve refutation. The truthfulness of the Pentateuch stands upon far higher grounds than any arithmetical tests can furnish. Upon this side of the argument I am precluded from entering, from the nature of my treatise. One thing is evident, that this attack, which receives all its importance from the position in the Church which the assailant occupies, is being made the occasion, and will be made more so, of enlarging and deepening, and not of weakening, the foundations upon which the arguments for the truth and inspiration of Holy Scripture rest.*

* I have been told that there are persons who scoff at the idea of the future resurrection, on the ground, that there would not be standing-room on the earth for all the persons who have been born since the creation of Adam. I have nowhere seen this in print, and therefore confine my remarks on the subject to a note. Were we unable to meet this objection by calculation, we might well assume, that on so unprecedented an occasion, in which the ordinary circumstances of the present order of things can be no guide to us, the Almighty will provide for the carrying out of His own purposes. But calculation shows the utter groundlessness of such an objection. I here give an outline which the algebraist can fill up. I will approximate to the total number of souls which have been born since the days of Noah, when the population of the world started afresh from eight persons : it may now be estimated at one thousand millions. There have been 126 generations in this time, reckoning three to a century. The generations from Adam to Noah were far fewer, and therefore the number of souls which my calculation will bring out will be very far

320 ARGUMENT FROM PAST EXPERIENCE.

I have now closed my Examples. Four of them appertain to a time prior to the development of modern science; and the remaining twenty-one fall within that period. Our experience, then, of the real harmony of Scripture and Science, even in cases where that harmony has for a time been supposed by some not to exist, ranges over many centuries, and reaches down to our own days.

Argument from past experience summed up.

With so many examples before us, I think we may boldly say, that even to suspect that Scripture and

---

more than half the whole number of souls which have been born since Adam.

Let b be the ratio of the average number of births during one generation to the population at the beginning of that generation. i the ratio of the average increase of population during a generation to the population at the beginning of the generation.

In order to meet extraordinary casualties, arising from war, pestilence, famine, and floods, I will suppose that, at the end of each generation, the population is reduced in the ratio of $1 : n$.

The algebraist will easily see that the population at the end of the 126th generation, that is, at the present time, equals

$$8 (1 + i)^{126} n^{125}, \text{ and this} = 10^9 \text{ by hypothesis.}$$

By reduction this becomes

$$(1 + i) n = 1 \cdot 16 n^{\frac{1}{126}} \text{ nearly} = 1 \cdot 16 \text{ nearly};$$

for if we take the extravagant case of even half the population being swept away by extraordinary events, at the end of every generation, the result would be $1 \cdot 154$, which differs very slightly from $1 \cdot 16$.

Again, the algebraist will easily see that the total number of souls since the days of Noah equals

$$8b \frac{(1+i)^{126} n^{126} - 1}{(1 + i) n - 1}$$

By substituting the values I have found above this

$$= \frac{10^9 \, nb}{0 \cdot 16} \text{ very nearly.}$$

Science can ever be opposed to each other is UNPHILOSOPHICAL. With such experience as the past has heaped up for our instruction and warning, is it not in the highest degree contrary to the spirit of true philosophy to sound the alarm at every appearance of antagonism between the Word and Works of God? Have not the scientific, in the steady advance of truth, been forced times without number to abandon theories which once appeared plausible and comprehensive, but which could not satisfy the stern requirements of fact? Have anomalies and contrarieties staggered them, and not rather quickened their search for clearer light and a nearer acquaintance with hidden connexions? And why should not the same waiting and trustful spirit

---

The radius of the earth is about 3956 miles. Suppose that only one-third of the surface is land, and that s square yards is the average space which each individual, of all who have been born from Noah downwards, would have for standing upon. It can easily be deduced, that the number of persons who could be accommodated

$$= \frac{20459}{s} \times 10^{10}$$

Put this equal to the number of births; then

$$s = \frac{32734}{nb}, \text{ which is larger than } \frac{32734}{b},$$

because n is necessarily smaller than unity. The value of s depends upon b, or the ratio of the number of births in one generation to the population at the beginning of the generation. It is hardly an over-estimate to say, that under ordinary circumstances the population will double in one generation of 33 years, that is, $i = 1$; and that the number of births will not more than double the number of deaths during a generation; that is, that b will not be greater than 2. Hence s will not be less than 16367 square yards, or a square space 128 yards each side! ample room; this is about 1900 persons to a square mile.

guide us when the Holy Scriptures are involved; especially when we remember the trophies of victory they bring with them from so many previous conflicts? With the history of past conflicts and past triumphs before us, whatever startling difficulties may yet arise, we shall do well to pause and wait for further light ere we venture to assert, or even to suspect, that they are enemies whom we have found to be friends under so many trying circumstances. Let our inductions be sober and well weighed, and we need have no fear that the Scriptures, as the inspired Word of God, and Science, as the means of setting forth the glory of His works, will ever be found not to speak the same thing in matters which they touch upon in common.

The last example of my first set, that referring to the motion of the earth, appertains to the sixteenth century. At that epoch the human mind in civilized Europe underwent a marked change, and great philological and scientific light dawned upon it. The long-standing dogma of the immobility of the earth was swept away by the unanswerable arguments which scientific reasoning produced; and the new principles of exegesis, which the revival of literature brought in, have removed also the difficulties which Scripture appeared to present to the reception of the new views. The human mind, however, appears to some to be, at the present day, going through another crisis: and it has been suggested, that as a change took place in the principles of interpretation on that former occasion, so another may be expected now; and, therefore, that the analogy which I have drawn between the *past*, in which many apparent discrepancies between Scripture and

Science have been removed, and the *future*, which may produce more, may not hold.*

(1.) On this I remark, in the first place, that twenty-one of my twenty-five examples, in which the charge that Scripture and Science are at variance has been disproved, are taken from modern discoveries, and from questions which are even now, or till very recently have been, debated: so that they belong to this era—an era which is well characterized as being one of great scientific activity and mental restlessness. The foundation, therefore, of my argument, for confidence in a never-ceasing harmony existing between Scripture and Science, whatever new discoveries may arise, lies as much in the experience of the present movement as in that of any earlier era of scientific progress; indeed I may say that my illustrations are far more drawn from the present period than from the earlier; and therefore there is no reason for doubting that the analogy will hold during the further development of modern discoveries.

(2.) I would next observe that there is a very striking resemblance between the character of the present movement and that of the sixteenth century; and, therefore, whatever general experience we may deduce from the earlier, might *à priori* fairly be expected to hold good during the later period. Freedom of inquiry, independence of mere authority as such, an appeal to the facts of nature, the careful study of Scripture now laid open, a getting out of the old grooves in which men's minds had been moving for ages, and the searching after truth for truth's sake,—these are some of the charac-

* See *Guardian*, June, 1861, p. 537.

teristics of the earlier era, and they are equally so of the later or present era. If, from this point of view, these eras differ in any respect, I should say that it is to the decided disparagement of the present in one important particular, though, as regards activity, the means of observation, and the methodical accumulation of facts, it greatly surpasses every period which has preceded it. But in the indulgence of a spirit of hasty generalization, and in the habit of building up theories upon insufficient data, the present age is, in too many instances, a most unworthy successor of that of which I would take Newton, the pattern of all sound philosophers, to be the type. As far as the present age departs from the principles of inductive philosophy, which inculcates the most patient inquiry, and a total abstinence from the use of hypothesis, except with the utmost caution, and upon a *vera causa* being shown, so far will its activity carry it away into the uncertain regions of speculation; and truth, the way to which is only along the roadway of facts, will never be reached, or, if reached, not known to be truth.

(3.) There is, however, one point in which the present era differs considerably from the former one. A new school of historical criticism has sprung up in the interval between them; and it is possible that some may think that the gordian knot of difficulty and discrepancy may be cut at once, by applying the same principles to the earlier portions of Scripture, which have been applied to the earlier chapters of Roman history, before the invasion of Pyrrhus. Niebuhr tells us, in the Preface to his *History of Rome*, that Roman history was treated, during the first two centuries after the revival of letters, with the same prostration of the understanding and judgment to the written letter that had been handed down, and with the same fearfulness of going beyond it, which prevailed in all the other branches of knowledge. From a multitude of insulated details writers drew up, what the remains of ancient literature did not give united in any single work, a systematic account of Roman antiquities. In the way of history, strictly so called, little was produced. In the latter half of the seventeenth century, philosophy entered upon a new state, for which she was to be indebted to the development of other sciences. Men were taught by great examples to look things in the face, and to pursue their researches with freedom; to regard the books, which till then had made the scholar's whole world, as merely pictures of a part of the living universe, which could not be directly approached; to exercise their own understanding, their own reason, their own judgment in everything. The end of the last century was the opening of a still more advanced era for Germany. 'We had now,' he writes, 'a literature worthy of our nation and language. . . . For this advantage Germany is indebted to Voss . . . with whom a new age for the knowledge of antiquity begins. Previous ages had been content to look at the ancient historians in the way many look at maps or landscapes, as if they were all in all; without ever attempting to employ them as the only remaining means for producing an image of the objects they represent. But now a work on such subjects could not be esteemed satisfactory, unless its clearness and distinctness enabled it to take its stand beside the history of the present age. . . . In this manner the critical treatment of Roman history, the discovery of the forms of the constitution which had till then been misunderstood, was a fruit that time had been maturing.'* Niebuhr accordingly writes his history from this point of view; very reasonably disallowing the 'marvellous,'† as inconsistent with the history he was endeavouring to trace; not suffering Livy's apparent credulity, when he would not absolutely reject 'those primitive ages when the gods walkt about among mankind;' and treating 'the heroes and patriots of Rome . . . not like Milton's angels, but as beings of our own flesh and blood.'‡ This free handling and rational method is, of course, the only one likely to extract anything like truth from mere 'fables and disfigured legends' which carry with them no pretence to history, much less to inspiration.

The same method of proceeding, however, is not altogether applicable to the Sacred Scriptures. As far as the rules of ordinary criticism go, they should be applied, no doubt, to the language of the Sacred Volume as to any other book; but not to its contents. I write for those who believe in the possibility of miracles, and not for those who deny it. The book, and many of the events and statements recorded in it, are avowedly of superhuman origin. Miracles are not matters of reasoning, but of *testimony*. The point which we have to ascertain on taking up this book is, whether it is genuine and authentic. This being determined, the superhuman character of many of its

* Preface to *Niebuhr's History of Rome*, translated by Hare and Thirlwall.
† *History*, p. 222.   ‡ Ibid. p. 3, 4.

statements and events must be received and implicitly believed in: for they do not come within the category of ordinary events. To go over the vast field of evidence of the credibility and authenticity of the Sacred Volume would be out of place in this treatise. But I may here bring together the numerous indications which I have adduced in the foregoing pages, that the tendency of modern scientific investigations— philological, ethnological, and antiquarian—is all in favour of the truthfulness of the Scripture records, even the most ancient. Not only have the exhumed remains of Nineveh and Babylon testified—for example, the remarkable discovery of the name Belshazzar in the inscriptions (see p. 166)—to the truth of Scripture history, at a date as early as any well-authenticated profane history; but in mounting up the stream of time beyond that date, wherever any glimpse can be caught from contemporary evidence of any kind, it bears testimony to the truth of the Scripture narrative. Thus, for example, the only point of contact which Egyptologists profess to have discerned in the hieroglyphic inscriptions between Egypt and Palestine bears witness, as far as it goes, to the fact related in 2 Chron. xii., that Shishak, king of Egypt, came up against Jerusalem and took it (see p. 173). The discovery of the remains of cities in the Hauran, according precisely as they do both in character and position with those described in Deuteronomy as the cities of Bashan, testifies in like manner to the accuracy of the Scripture narrative at a still earlier epoch (see p. 166). The account of wars between Israel and Moab related in 2 Kings, iii. 4-27, and 2 Chron. xx. is confirmed by the recently discovered Moabitic inscription (see p. 166). The original unity of language, and its violent separation into distinct languages related in the eleventh chapter of Genesis, and the testimony borne to both these facts by the discoveries of philologists (see p. 142) —the remarkable coincidences which are traceable between the statements of the tenth chapter of Genesis, and the discoveries of ethnologists, and also the known geographical distribution of nations (see p. 145)—the concurrent testimony borne by many and widely-scattered nations, to the fact of a deluge, universal as regards the human race, having occurred in ancient times, and the correspondence between this tradition and the narrative of the flood, related as part of a continuous history in the sixth, seventh, and eighth chapters of Genesis—the testimony of the best physiologists that all human races are of one species, and the agreement of this with the account of the creation of Adam and Eve, in the first and second chapters of Genesis, and the descent of all men from them (p. 104), —all these comparisons point in the same direction, and have a cumulative effect in confirming our belief that even the most ancient portion of Scripture is an historical document. In corroboration of this, I may add that no arguments on the opposite side stand their ground, whether drawn from Egyptian antiquities, Nile deposits, or Hindoo or Chinese astronomical observations, or from any other source.

I may also observe, that the composition of the earlier chapters of Genesis seems to bear no such analogy to that of the earlier chapters of Roman history, as to call for the handling which these latter have received, and so to be deprived of their character as simple history. 'Although there is a continuity of narrative,' says Sir G. C. Lewis, 'running through the story of Romulus; though the successive events stand to one another in an intelligible relation of cause and effect, yet we can trace the deliberate invention of the ætiologist. The story is formed by an aggregation of parts: there is no uninterrupted poetical flow, or epic unity. Instead of resembling a statue cast in one piece in a foundry, it is like a tessellated pavement, formed into a pattern by stones of different colours.'* Nothing of artifice of this kind can be traced in the earlier chapters of Genesis. After the narrative of the Creation and Fall of Man in the first three chapters, there are only two—the fourth and fifth—before the narrative of the flood begins. This brings us down to the tenth, which describes the nations as 'divided in the earth after the flood,' and then the eleventh, relating to the confusion of tongues, and ending with a genealogy from Noah to Abram. From that point the history of the chosen race begins, and runs through the whole of Scripture to the end of the Old Testament, with the single exception of the Book of Job. The fourth and fifth chapters, which I have passed over, give a variety of isolated facts regarding certain individuals, from Cain and Abel down to Lamech, and the birth of his son Noah and his grandsons Shem, Ham, and Japheth, together with the intervening genealogy. Now, all this, no doubt, may be compared, in one sense,

* *An Inquiry into the Credibility of the Early Roman History*, vol. i. p. 437.

to a tessellated pavement. But there is no attempt at moulding the parts into one another. The whole appears to be a series of matter-of-fact statements, without any decoration. No ulterior end appears to have been in view, such as in the Roman story, to give a plausible account of the origin of institutions and customs which had subsequently grown up in the national history. The formal incorporation of the Sabbath in the moral law, delivered on Mount Sinai, is the only apparent instance of this, as the argument for it is based upon the six days of creation and the seventh of cessation from work. But the Sabbath had existed before the giving of the law on Mount Sinai, as is evident from the regulations regarding the gathering of the manna (Exod. xvi. 22, 23) and from other reasons. None of the Jewish institutions, in fact, are to be traced to the events narrated in these first eleven chapters of Genesis. Indeed, more allusions to these events occur in the New Testament than in the Old. There appears, then, to be no valid reason for the application of the new principle of historical criticism to the early history of the Book of Genesis.

The inference, that because it is found in various instances that the dawn of genuine history in civilized nations is preceded by a series of myths and legends, therefore the same must be the case with the Hebrew nation and the documents they have handed down to us, is a gratuitous assumption and an unphilosophical begging of the whole question. It would be as fair to infer that if ten counterfeit coins were thrown into circulation together with the genuine coin of the realm, and by our sagacity the counterfeit coins were detected by us, one by one, the pure coin must be declared to be counterfeit also. Each case must stand upon its own merits. Indeed, the earliest records of other nations, as is acknowledged by those critics, are uniformly devoted to the suspicious task of tracing their own origin and exalting their own early glories. But the earliest of the documents the Hebrews have handed down to us in the sacred Scriptures betray nothing of this kind: they make no peculiar reference to the Jewish nation, but appertain solely to the human race at large. Nor in those subsequent parts of these documents, which relate the rise of their own nation in the call and history of Abram, is there any undue glorification of their great progenitor; but the blemishes of his character are unsparingly recorded in a manner in which no feigned historian would ever venture to indulge.

In these remarks I have, of course, been regarding Scripture as I would any other book, the historical credibility of which was under trial. The testimony of contemporary history utterly fails in this case, as no history and no monuments can pretend to compete with the Book of Genesis in point of antiquity. Such arguments of a secular character as were available, in the absence of this testimony, I have endeavoured to bring to bear upon the subject. But while I altogether concur in the propriety of keeping the scientific and the religious sides of the question entirely apart during the discussion, I think that for any one, in the pursuit of truth, altogether to ignore in such a case the religious argument, is as preposterous as if he were to agree to argue regarding the doctrines of Christianity upon the hypothesis that Christ was a mere man and not the Son of God. The witness which Christ gives to the historical character of Genesis, I reserve to the next Part. His witness is true; and, in the absence of collateral testimony, is invaluable.

I have been arguing on these questions as if they were quite isolated, and were to be tested by the bare probabilities of abstract reasoning, without regard to that later revelation, under which Science has grown to be what it is. But it is clearly unreasonable to *isolate* the argument thus. The Gospel throws rays of glory backward as well as forward. The Book of Genesis partakes of the radiance. The first eleven chapters of the book, even as a fragment, are of the deepest value, as I shall presently show; but their full grandeur is apparent only when viewed as a part of the unique volume at the head of which they stand.

PART II.

THE HISTORICAL CHARACTER, PLENARY INSPIRATION, AND SURPASSING IMPORTANCE OF THE FIRST ELEVEN CHAPTERS OF GENESIS.

In the First Part the argument has been rather of the negative description. The high *improbability* of Scripture being at variance with Science has been established; and that, not so much from a consideration of the character of Scripture itself, as from the experience of the past; which shows so many instances of imaginary discrepancies becoming in the end witnesses on the other side, and illustrating with such force the harmony between the Word and Works of God, that any man who ventures to set aside this experience justly forfeits the title of Philosopher.

In the present part of my treatise I propose to make some remarks on the character and contents of the earlier portion of the sacred volume, selecting for this purpose the First Eleven Chapters of Genesis, as it is here that Scripture and Science have been supposed more particularly to come into collision. On the historical character of Scripture some remarks have already been made in the last few pages. I hope now to establish the historical character and plenary inspiration of this portion of the Sacred Volume by a reference to the language of our Lord and His Apostles, and to point out in various eminent particulars its surpassing importance. An argument of a positive nature, and confirmatory of that which I have already given, will thus flow from the character of Scripture itself, to show how *impossible* it is that such a record can in any way contravene the teachings of the phenomena and laws of the material world which proceeds from the same almighty Author. This part of my treatise is expressly addressed to Christians, who regard the testimony of our Lord and His Apostles as infallible; not to men who, because our Lord '*increased in wisdom*' as well as 'stature,' consider it to be 'perfectly consistent with the most entire and sincere belief in our Lord's Divinity,' to hold that our Lord may have asserted that which was not true, if it accorded with the views of the most learned of His nation! In this way such men endeavour to get rid of the testimony of these texts; John, v. 46, 47; Luke, xvi. 29, 31; xx. 37.*

* Colenso's *Pentateuch*, Part I. p. xxx., xxxi.

CHAPTER I.

THE HISTORICAL CHARACTER AND PLENARY INSPIRATION OF THE FIRST ELEVEN CHAPTERS OF GENESIS.

BY the inspiration of Holy Scripture I understand, that the Scriptures were written under the guidance of the Holy Spirit, who communicated to the writers facts before unknown, directed them in the selection of other facts already known, and preserved them from error of every kind in the records they made. Definition of Inspiration.

This definition I have given in every edition of my book. Dr. Colenso quotes it in his attack upon the historical character of the Pentateuch, printing the last clause in italics, as the text from which he proceeds in his investigation. It is the inspiration of the original documents which have long ago disappeared that I contend for. That errors have crept into the long succession of copies no one denies. Had Dr. Colenso, by his arithmetical test, detected any internal contradictions, all it would have amounted to would be, that our present copies have become, in those respects, more corrupted by transmission from hand to hand than was before known. But that the documents are unhistorical, or comparatively modern productions not to be depended upon as history, would by no means be proved by such a process. Dr. Colenso's criticisms, as objections to the historical character of the Pentateuch, have been shown in a former page to be without any value.

A friendly reviewer* of my fifth edition also thinks, that, by claiming freedom from error of every kind for the originals of Holy Scripture, I have made 'an assumption which is absolutely gratuitous, and which imposes a formidable and needless task upon the defenders of inspiration.' But to this I reply, that if *all scripture is given by inspiration of God* (2 Tim. iii. 16), and if this, as the context proves,† applies, not only to parts of scripture, but to all scripture, the matter is settled; for it is not to be allowed for one moment, that the Holy Spirit would either dictate or suffer error of any kind in what was written under His suggestion or superintendence. If, on the other hand, we suppose that some parts are inspired and other parts not, where is the line to be drawn? We must take our standpoint somewhere. It has always appeared to me most reasonable, and most consonant with our ideas of the Divine Being, to take up the position that Holy Scripture, coming as it does from God, is *perfect*, till the contrary has been proved; and to consider that the discrepancies among MSS. are to be attributed to the numerous host of transcribers through whose hands our present copies are come down to us—discrepancies which may be counted by thousands, but which, by the acknowledgment of those who have hunted for contradictions, are of no importance and are what can

---

* *Guardian*, August 30, 1865.
† See Note on page 347.

readily be accounted for by the natural fallibility of the most careful copyist. The argument of the sceptic, against those who admit of only a partial inspiration of the Holy Volume, is to me unanswerable, and represents the views I have always held on the subject. He says, 'A book cannot be said to be inspired, or to carry with it the authority of being God's Word, if *only portions* come from Him, and there exists no plain and infallible sign to indicate *which* those portions are; and if the same writer may give us in one verse of the Bible *a revelation from the Most High*, and in the next verse *a blunder of his own*. How can we be certain, that the very texts, upon which we rest our *doctrines* and our *hopes*, are not the *uninspired* portion? What can be the meaning or nature of an Inspiration to teach Truth, which does not guarantee its recipient from teaching error?'*

I would make every concession which a fair criticism of the text will allow: but to go further than this I think is dangerous. To place ourselves on the same low level with those who treat portions of Holy Scripture as fable, or, at any rate, as simply human, and to contend with them with their own weapons, as many attempt to do, is forsaking that high vantage ground which a believer in Revelation can claim, and is trusting too much to our own prowess. To stand on the defensive, and challenge proof of the works of God in any single instance being at variance with His word, is the safest and the right attitude for us to assume.

---

* Quoted from Bishop Wordsworth's *Lectures on the Inspiration of the Bible*, in which the above words are 'transcribed from a volume recently published by a sceptical writer.' Seventh Edit. p. 11, 12.

If we descend, we can do so only by making compromises: we must waive the miraculous and the divine, we must use arguments drawn in every case only from the ordinary course of things. Thus Dr. Colenso wonders (p. 36) that I pass over the great event in Joshua's history so lightly, as he regards it as the most striking of instances of Scripture and Science being at variance: whereas I regard it as a Divine interposition, a miracle, and therefore as an event not coming within the range of scientific investigation. So, in his remarks upon the deliverance of the Israelites from Egyptian bondage, he appears to ignore altogether the Divine help in the whole transaction. A certain Greek professor once said that the Bible should be treated critically like any other book—to which principle I would agree when applied properly. But I altogether dissent, when he applies his principle thus; that, as Isaiah mentions the name of Cyrus, he must have lived after Cyrus. And why do I dissent? Because the book of Isaiah is avowedly prophetical: and should therefore be tested by other means, unless it can be proved that prophecy is impossible, which is out of the question. My object has not been to reconcile Scripture and Science (though I think this is generally the result of my treatise), but to show that Science cannot *prove* anything to be *at variance* with Scripture: and in doing this I have taken the highest ground, and assumed that Scripture, in its originals, is free from error of every kind: and I think I have thus far maintained it.

I now proceed to show further, that the first eleven chapters of Genesis, the portion of Scripture now specially under consideration, are a genuine historical document and must come under my definition of Inspiration. This I do by an appeal (1) to the use made of them by our Lord and His Apostles, (2) to the matters which they contain, and (3) to the early place they occupy in the Sacred Volume.

1. The following series of verses chosen from these chapters, with corresponding texts from the New Testament placed after them, shows how repeatedly this portion of Scripture is either quoted or referred to by our Lord and His Apostles:—

Gen. I. 1. In the beginning God created the heaven and the earth.
Heb. i. 10 (quoting *Ps* cii.).—Thou, Lord, in the beginning hast laid the foundation of the earth, and the heavens are the works of thine hands.
Heb. xi. 3.—Through faith we understand that the worlds were framed by the word of God, so that things which are seen were not made of things which do appear.

3. And God said, Let there be light: and there was light.
2 *Cor.* iv. 6.—God, who commanded the light to shine out of darkness.

9. And God said, Let the waters under the heaven be gathered together unto one place, and let the dry land appear: and it was so.
2 *Pet.* iii. 5.—By the word of God the heavens were of old, and the earth standing out of the water and in the water.

11. And God said, Let the earth bring forth grass, the herb yielding seed, and the fruit tree yielding fruit after his kind, whose seed *is* in itself, upon the earth: and it was so.
Heb. vi. 7.—The earth which . . . bringeth forth herbs meet for them by whom it is dressed, receiveth blessing from God.

26. And God said, Let us make man in our own image, after our likeness: and let them have dominion over the fish of the sea, and over the fowl of the air, and over the cattle, and over all the earth, and over every creeping thing that creepeth upon the earth.
Heb. ii. 7, 8 (quoting *Ps.* viii. 6).—Thou madest him a little lower than the angels; thou crownedst him with glory and honour, and didst set him over the works of thy hands: thou

hast put all things in subjection under his feet.

27. So God created man in his *own* image, in the image of God created he him: male and female created he them.

1 *Cor.* xi. 7.—For a man indeed ought not to cover his head, forasmuch as he is the image and glory of God.

James, iii. 9.— . . . . men, which are made after the similitude of God.

Matt. xix. 4.—Have ye not read, that he which made them at the beginning made them male and female?

II. 1. Thus the heavens and the earth were finished, and all the host of them.

Acts, iv. 24.—Lord, thou art God, which hast made heaven, and earth, and the sea, and all that in them is.

Acts, xiv. 15.— . . . the living God, which made heaven, and earth, and the sea, and all things that are therein.

Acts, xvii. 24.—God that made the world and all things therein.

Eph. iii. 9.— . . . God, who created all things by Jesus Christ.

Col. i. 16.—For by him were all things created, that are in heaven, and that are in earth, visible and invisible, whether they be thrones, or dominions, or principalities, or powers: all things were created by him and for him.

Heb. i. 2.— . . . by whom also he made the worlds.

Heb. iii. 4.—He that built all things is God.

Rev. iv. 11.— . . . for thou hast created all things, and for thy pleasure they are and were created.

Rev. x. 6.—And sware by him that liveth for ever and ever, who created heaven, and the things that therein are, and the earth, and the things that therein are, and the **sea**, and the things which are therein.

Rev. xiv. 7.—Worship him that made heaven, and earth, and the sea, and the fountains of waters.

2. And on the seventh day God ended his work which he had made; and he rested on the seventh day from all his work which he had made.

Heb. iv. 4.—For he spake in a certain place of the seventh day in this wise, And God did rest the seventh day from all his works.

3. And God blessed the seventh day, and sanctified it: because that in it he had rested from all his work which God created and made.

Mark, ii. 27.—The sabbath was made for man, and not man for the sabbath.

7. And the Lord God formed man *of* the dust of the ground, and breathed into his nostrils the breath of life; and man became a living soul.

Luke, iii. 38.—Adam, which was the son of God.

1 *Cor.* xv. 45.—And so it is written, The first man Adam was made a living soul.

1 *Cor.* xv. 47.—The first man is of the earth, earthy.

9. And out of the ground made the Lord God to grow every tree that is pleasant to the sight and good for food; the tree of life also in the midst of the garden, and the tree of knowledge of good and evil.

Rev. ii. 7.—To him that overcometh will I give to eat of the tree of life, which is in the midst of the Paradise of God.

18. And the Lord God said, *It is* not good that the man should be alone; I will make an help meet for him.

1 *Cor.* xi. 9.—Neither was the man created for the woman; but the woman for the man.
1 *Tim.* ii. 13.—For Adam was first formed, then Eve.

23. And Adam said, This *is* now bone of my bones, and flesh of my flesh: she shall be called Woman, because she was taken out of Man.

1 *Cor.* xi. 8.—For the man is not of the woman; but the woman of the man.

24. Therefore shall a man leave his father and mother, and shall cleave unto his wife: and they shall be one flesh.

Matt. xix. 4, 5.—And he answered and said unto them, Have ye not read, that he which made them at the beginning made them male and female, and said, For this cause shall a man leave father and mother, and shall cleave to his wife: and they twain shall be one flesh?—*See also Mark,* x. 6-8.
Rom. vii. 2.—The woman which hath an husband is bound by the law to her husband so long as he liveth.
1 *Cor.* vi. 16.— . . . . for two, saith he, shall be one flesh.
Eph. v. 30, 31.—For we are members of his body, of his flesh, and of his bones. For this cause shall a man leave his father and mother, and shall be joined unto his wife, and they two shall be one flesh.

III. 4. And the serpent said unto the woman, Ye shall not surely die:
5. For God doth know, that in the day ye eat thereof, then your eyes shall be opened, and ye shall be as gods, knowing good and evil.

John, viii. 44.—Ye are of your father the devil. . . . He was a murderer from the beginning, and abode not in the truth, because there is no truth in him.
2 *Cor.* xi. 3.—But I fear, lest by any means, as the serpent beguiled Eve through his subtlety, so your minds should be corrupted from the simplicity that is in Christ.
1 *John,* iii. 8.—He that committeth sin is of the devil; for the devil sinneth from the beginning.
Rev. xii. 9.—And the great dragon was cast out, that old serpent, called the Devil, and Satan, which deceiveth the whole world.—*See also* xx. 2.

6. And when the woman saw that the tree *was* good for food, and that it *was* pleasant to the eyes, and a tree to be desired to make *one* wise, she took of the fruit thereof, and did eat, and gave also unto her husband with her; and he did eat.

1 *Tim.* ii. 14.—And Adam was not deceived, but the woman being deceived was in the transgression.

15. And I will put enmity between thee and the woman, and between thy seed and her seed; it shall bruise thy head, and thou shalt bruise his heel.

Rom. xvi. 20.—And the God of peace shall bruise Satan under your feet shortly.

16. Unto the woman he said, I will greatly multiply thy sorrow and thy conception; in sorrow thou shalt bring forth children; and thy desire *shall be* to thy husband, and he shall rule over thee.

1 *Cor.* xi. 3.— . . . the head of the woman is the man.

1 *Cor.* xiv. 34.— . . . they are commanded to be under obedience, as also saith the law.

17. And unto Adam he said, Because thou hast hearkened unto the voice of thy wife, and hast eaten of the tree, of which I commanded thee, saying, Thou shalt not eat of it; cursed *is* the ground for thy sake; in sorrow shalt thou eat *of* it all the days of thy life.

18. Thorns also and thistles shall it bring forth to thee; and thou shalt eat the herb of the field:

19. In the sweat of thy face shalt thou eat bread, till thou return unto the ground; for out of it thou wast taken: for dust thou *art,* and unto dust shalt thou return.

Rom. v. 12.— . . ., by one man sin entered into the world, and death by sin.

Rom. vi. 23.—The wages of sin is death.

Rom. viii. 20, 22.— . . . The creature was made subject to vanity. . . . the whole creation groaneth.

1 *Cor.* xv. 21, 22.—By man came death. . . . in Adam all die.

Heb. ix. 27.—It is appointed unto men once to die.

20. And Adam called his wife's name Eve; because she was the mother of all living.

Acts, xvii. 26.—And hath made of one blood all nations of men.

IV. 4. And Abel, he also brought of the firstlings of his flock and of the fat thereof. And the Lord had respect unto Abel and to his offering.

5. But unto Cain and to his offering he had not respect.

Heb. xi. 4.—By faith Abel offered unto God a more excellent sacrifice than Cain, by which he obtained witness that he was righteous, God testifying of his gifts.

8. And Cain talked with Abel his brother: and it came to pass, when they were in the field, that Cain rose up against Abel his brother, and slew him.

Matt. xxiii. 35.— . . . all the righteous blood shed upon the earth, from the blood of righteous Abel. . . . —*See also Luke* xi. 51.

1 *John,* iii. 12.—Not as Cain, who was of that wicked one, and slew his brother.

Jude, 11.—They have gone in the way of Cain.

10. And he said, What hast thou done? the voice of thy brother's blood crieth unto me from the ground.

Heb. xii. 24.— . . . the blood of sprinkling, that speaketh better things than that of Abel.

V. 3. And Adam begat *a son* . . . and called his son Seth. 6. And Seth . . . begat Enos. 9. And Enos . . . begat Cainan, &c. . . . 32. And Noah begat Shem.

Luke, iii. 36-38.— . . . which was the son of Sem, which was the son of Noe. . . . which was the son of Adam.
Jude, 14.—Enoch, . . . the seventh from Adam.

---

24. And Enoch walked with God: and he *was* not; for God took him.

Heb. xi. 5.—By faith Enoch was translated that he should not see death; and was not found, because God had translated him.

VI. 3. And the Lord said, My spirit shall not always strive with man, for that he also *is* flesh: yet his days shall be an hundred and twenty years. . . 14. Make thee an ark of gopher wood . . . 17. And behold, I, even I, do bring a flood of waters upon the earth, to destroy all flesh . . . 18. But . . . thou shalt come into the ark, thou, and thy sons, and thy wife, and thy sons' wives with thee. 22. Thus did Noah.

Heb. xi. 7.—By faith Noah being warned of God of things not seen as yet, moved with fear, prepared an ark to the saving of his house.
1 *Pet.* iii. 20.—The long-suffering of God waited in the days of Noah, while the ark was a preparing.

VII. 7. And Noah went in . . . into the ark . . . 17. And the waters increased, and bare up the ark . . . 21. And all flesh died that moved upon the earth . . . 23. . . . and Noah only remained *alive,* and they that *were* with him in the ark.

Luke, xvii. 27.—Noe entered into the ark, and the flood came, and destroyed them all.—*See also Matt.* xxiv. 38.
1 *Pet.* iii. 20.— . . . the ark . . . wherein few, that is, eight souls, were saved by water.
2 *Pet.* ii. 5.—And spared not the old world, but saved Noah the eighth person, a preacher of righteousness, bringing in the flood upon the world of the ungodly.
2 *Pet.* iii. 6.—Whereby the world that then was, being overflowed with water, perished.

IX. 3. Every moving thing that liveth shall be meat for you; even as the green herb have I given you all things.

1 *Tim.* iv. 3.— . . . meats, which God hath created to be received with thanksgiving of them which believe and know the truth.

---

4. But flesh with the life

thereof, *which* is the blood thereof, shall ye not eat.
_{*Acts*, xv. 29.—That ye abstain . . . from blood.}

11. . . . neither shall all flesh be cut off any more by the waters of a flood; neither shall there any more be a flood to destroy the earth. *See also* VIII. 22.
_{2 *Pet*. iii. 7.—But the heavens and the earth, which are now, by the same word are kept in store, reserved unto fire against the day of judgment.}

X. 32. These *are* the families of the sons of Noah, after their generations, in their nations: and by these were the nations divided in the earth after the flood.

_{*Acts*, xvii. 26.—And hath made of one blood all nations of men for to dwell on all the face of the earth, and hath determined the time appointed, and the bounds of their habitations.}

XI. 10. Shem . . . begat Arphaxad . . . 12. And Arphaxad begat Salah, &c. . . . 26. And Terah . . . begat Abram.
_{*Luke*, iii. 34-36.—Abraham, which was the son of Thara . . . which was the son of Sem.}

31. And Terah took Abram his son . . . and they went forth . . . from Ur of the Chaldees . . . and they came unto Haran, and dwelt there.
_{*Acts*, vii. 2.—Abraham, when he was in Mesopotamia, before he dwelt in Charan.}

1. Here are sixty-six passages of the New Testament in which these eleven chapters of Genesis are either directly quoted, or are made a ground of argument. Of these, six are by our Lord Himself, two of them being direct quotations; thirty-eight by St. Paul, three being direct quotations; six by St. Peter; seven in St. John's writings; one by St. James; two by St. Jude; two by the assembled Apostles; three by St. Luke, two of them direct quotations; and one by St. Stephen.

The inference which I would draw from this circumstance is, that our Lord and His Apostles regarded these eleven chapters as *historical documents worthy of credit*, and that they made use of them to establish truths—a

_{Our Lord and His Apostles regarded theseChapters as historical documents.}

thing they never would have done, had they not known them to be authoritative.

The texts quoted or referred to, lie, moreover, scattered through the chapters in such a way, that no claim of authority can be set up for one part which cannot be equally demanded for every other. The creation of matter, the formation of the worlds and all things in them, the command that light should shine, the separation of the waters and the appearance of dry land; the creation of man out of the dust; dominion given him over the irrational world; his being made in the image of God, and having a living soul; God resting the seventh day from all His works, and the institution of the Sabbath; Adam being first made and then Eve, the woman from and for the man, and their being made male and female; the law of marriage, and the man cleaving unto his wife so that two should be one flesh; the Fall, and the entrance of sin and death into the world, its origin in Satan's guile tempting and deceiving Eve in the form of a serpent with lying words, Adam not deceived but tempted through Eve, the curse upon the serpent that the seed of the woman should bruise the serpent's head; the origin of all nations in one blood; the existence of the tree of life in Paradise; the history of Cain and Abel; the genealogy from Adam to Noah, and also from Noah to Abraham; the exalted character of Enoch; circumstances connected with the Deluge; and the subsequent division of the earth among the descendants of Noah;— all these topics are severally drawn from the chapters under consideration, by our Lord and His Apostles, in a manner which precludes every hypothesis, but that they regarded the opening portion of the Sacred Volume to be of historic credibility and plenary authority. The notion that these chapters are myths or merely parabolic, as some have hastily conjectured, or that the account of the Creation and Fall was acted over in a vision to Moses, as others have suggested, is only a device for getting rid of their historical character. But the fact that they are quoted and used as circumstantial narratives by infallible authority, can never be thus set aside.

2. The historical credibility of these chapters having been thus established, the proof of their inspiration Their Inspiration follows from their contents. immediately follows from an examination of their contents. Nothing short of a Divine communication could have furnished the information which they contain regarding the origin of matter, and the formation of the worlds; the origin of man, his original purity and uprightness, his soul made in the image of God; the history of the entrance of sin and death into the world; and the two Paradisaical institutions of the Sabbath and Marriage. While, then, the seal set to the document by our Lord and His Apostles stamps these communications as unerring verities, the nature of the truths themselves leaves no alternative but to conclude, that they were written under the inspiration of God.

3. These same transcendental truths appear in other parts of Scripture; and it is possible that, in treating These Chapters are not quotations. of them there, an objector to Inspiration might deny the inference which I have drawn. But here there is no room for such denial; here they appear for the first time. And, consequently, though it might be suggested by some

regarding other parts of Scripture containing them, that there the statements were borrowed, and that therefore those portions could not carry with them their own evidence of inspiration, any more than any ordinary human compositions quoting the same, here there can be no doubt. Here is the original source and fountain, from which all subsequent information regarding them is drawn. These are the very documents to which the prophets, from Moses downwards, and our Lord Himself and His Apostles at the end of that long period, pointed as a true source and fountain of information.

It is difficult to suggest what stronger evidence could have been given than this, short of a voice from heaven which every ear could hear in every age. But even then, no doubt, the hesitations of unbelief would darken the testimony; and it would still in a way be true, that 'if they hear not Moses and the prophets, neither will they be persuaded though one rose from the dead' (Matt. xvi. 31). A celebrated infidel once declared, that if Christianity were true, it would have been recorded in unmistakable characters upon the heavens, that none might have room to doubt in so important a matter. But he was met by the query, whether this was the case in natural religion—whether it is not by patient inquiry and diligent examination of evidence, which the works of God and our consciences supply, that the first principles of natural religion are discovered and believed in? So of Christianity: so of Inspiration. We must weigh the evidence: for it is upon evidence alone that we can determine.*

* In the First Edition (1856) I gave a sketch of the general argument for the inspiration of the whole Old Testament taken as one

It follows as a natural corollary from what has gone before, that in these chapters there is no admixture of

volume. I now remove those portions of the text to a note, and add some remarks on the inspiration of the New Testament also.

OUTLINE OF THE ARGUMENT FOR THE INSPIRATION OF THE OLD AND NEW TESTAMENTS.

The Old Testament consists of those writings which were known in the time of our Lord as αἱ γραφαί, or The Scriptures, and were admitted on all hands, by friends and enemies, to be of infallible authority, from which there was no appeal.

It is of these that St. Paul declares (2 Tim. iii. 16, 17), 'ALL SCRIPTURE IS GIVEN BY INSPIRATION OF GOD, and is profitable for doctrine, for reproof, for correction, for instruction in righteousness; that the man of God may be perfect, thoroughly furnished unto all good works.' And regarding the sacred penmen, St. Peter thus speaks (2 Pet. i. 21): ' Holy men of God spake AS THEY WERE MOVED BY THE HOLY GHOST.' And also the writer of the Epistle to the Hebrews (i. 1), ' God . . . at sundry times and in divers manners spake in times past unto the fathers by the prophets.'

Some have attempted to deprive us of the testimony of the first of these texts to the doctrine that the whole of Scripture is inspired, by suggesting that it might be differently translated in connexion with the context, and be made to say, not that every part of Scripture is inspired and profitable, but that all *those parts which are inspired* are ' profitable for doctrine, for reproof, for correction, for instruction in righteousness,' —a mere truism, compared with the lesson which our translation and, I feel persuaded, the original gives. They cannot deny that our own translation is a perfectly legitimate one; but they consider that theirs also is admissible. Apart, however, from philological and critical reasons, does this suit the context well ? St. Paul is commending the ' Holy Scriptures ' to Timothy, meaning the Old Testament in which he had been taught by his Jewish mother and grandmother; and he sums up his commendation with this text and what follows. If he meant to imply that only parts of these Holy Scriptures, which were received by the Jews and referred to by our Lord as a whole, were inspired, and that those were profitable, was it not a great omission, that he gave Timothy no canon by which he should know which were inspired, and which were not ? But no such canon appears, nor is one alluded to. In fact, this limiting hypothesis has no good argument to stand upon, and must be abandoned. The truth, however, of the

error of any kind, no inaccuracies arising from human ignorance and infirmity. For all is from God; and therefore bears His stamp. When the Maker of the world deigns to be- {These Chapters, therefore, contain no error.}

Inspiration of Holy Scripture as a whole does not rest solely on this text. If the text did not exist, the same result is to be gathered from other passages.
In entire accordance with these universal statements by St. Paul and St. Peter are the terms in which the Scriptures are incidentally referred to throughout the New Testament. See for example, Matt. i. 22; ii. 15. Mark, xii. 36. Luke, i. 70. Acts, i. 16; iii. 18; iv. 25; xxviii. 25. Rom. i. 2. Heb. iii. 7.
Take also the following, in which the Scriptures are spoken of in a way which belongs only to an Inspired Volume: Mark, vii. 13; John, x. 35; Acts, vii. 38: Rom. iii. 2; Eph. vi. 17; Heb. v. 12; 1 Pet. iv. 11.
The same Scriptures are continually referred to in the following terms:—' That the scripture might be fulfilled;' ' Have ye not read this scripture?' ' As the scripture saith;' ' What saith the scripture?' and similar forms of appeal occur in endless variety, all pointing to the sacred records as an infallible source of light and knowledge.
In strict keeping with this view are the declarations of the inspired penmen themselves: as the following, 2 Sam. xxiii. 1, 2; Isaiah, i. 10; Jer. i. 1, 2, and many such passages.
' Thus saith the Lord;' ' The mouth of the Lord hath spoken;' and many similar formulæ perpetually occur, all pointing to the same result—that the writers were conscious that they were speaking by the Spirit, and under His guidance.
But under the designation of ' Scripture' come, not only the Old Testament, but also those writings which have been handed down to us as the undoubted records of the Evangelists and Apostles, who were endowed by the Holy Ghost with all truth, as our Lord specially told them they should be, to teach and preach the Gospel, and to plant the Christian Church (See Matt. x. 19, 20; Mark, xiii. 11; Luke, xii. 11, 12; John, xiv. 16, 17, 25, and 26; xvi. 12–15). Can we for one moment suppose that they were thus inspired in their teaching, but not so in their writing?
That the term ' Scripture' is to be extended to the New Testament writings, is evident also from some incidental notices which lie in these writings, as seeds of evidence of the greatest value, now that the spirit

come an Author it is to be expected that His Word would be as perfect as His works. He is the God of

of scepticism would call their Inspiration in question. St. Peter speaks thus of St. Paul's writings. In alluding to some of the profound truths he delivered, he says,—' In which there are some things hard to be understood, which they that are unlearned [*i.e.* unlearned in the truth] wrest as they do also THE OTHER SCRIPTURES.' Hence St. Paul's writings came under the designation of 'Scripture,' in the times of the Apostles, as much as the Old Testament, here called 'the OTHER SCRIPTURES.' Again, St. Paul writes to Timothy, ' For the SCRIPTURE saith, Thou shalt not muzzle the ox that treadeth out the corn.' AND, ' The labourer is worthy of his reward.' (1 Tim. v. 18.) The first of these quotations is from Deuteronomy; but the second is found only in St. Luke (See x. 7; also, Matt. x. 10). Hence St. Paul calls this Gospel, which was then in circulation, the ' Scripture,' or part of the Scriptures.

Thus the New Testament, as well as the Old, comes under the designation of ' Scripture ;' and the whole volume must therefore, in every part of it, participate in this characteristic announced by St. Paul,—' ALL SCRIPTURE is given by inspiration of God.'

There are in this volume many things which reason could never have discovered; though there are other things, such as the acts of history, which needed no revelation. But if *all Scripture* is given by inspiration of God, we conclude that not only were those transcendental truths, for which the Bible is *unique*, revealed; but also, in all matters of an ordinary kind, the writers were directed by that same Spirit what to select, to form a part of the sacred deposit; and in their records were preserved from all error. Otherwise, I repeat, what can this mean? ' ALL SCRIPTURE is given by inspiration of God.'

Inspiration is a more general term than Revelation. Revelation applies solely to those things which could not be discovered by reason. Inspiration includes this, and more. It implies a Divine guidance in other things, when necessary. Aholiab and Bezaleel, who were raised up by God to construct the tabernacle in the wilderness, and all the curious carving for its decoration, were said to be ' filled with the Spirit ' of God for this work (Exod. xxxv. 31). ' The inspiration of the Almighty,' we are told, ' giveth understanding ' (Job, xxxii. 8). We pray for ourselves that God would ' cleanse the thoughts of our hearts by the inspiration of His Holy Spirit.'

We speak of *genius* sometimes as a kind of inspiration. We talk of the inspiration of the poet or of the musical composer. And some

truth, and is incapable of deceiving His creatures; it is impossible, therefore, that He should have introduced anything but truth into His revelation.

have attempted to account for the Inspiration which they cannot but acknowledge in some parts of the Scriptures, by referring it to a high-wrought description of genius. Genius is no doubt the gift of God; and is far more rarely bestowed than reason, another of His gifts. But it can never account for the presence of those lofty truths and those heaven-born revelations, which He only who is Infinite, to whom the past, present, and future, are all alike, could have communicated; nor can it account for those miraculous, as well as prophetic powers, with which many of the holy men of God were endowed, as a testimony to the truth of what they said and wrote.

The Inspiration, then, for which I am contending, cannot but be acknowledged in those parts of Scripture which have been obviously given by revelation. But Scripture, as a whole, is often appealed to as of final authority. How, then, can we admit that there is any admixture of human infirmity in any part? If there be error in any part, and this is not distinctly revealed to us, who is to draw the line, and say, 'This is true, and that doubtful?'

On the term ' Verbal Inspiration.'

The precise mode in which this power operated on the inspired writers is not told us; and, probably, were it revealed to us, it would be beyond our comprehension. Fanciful and overstrained notions have sometimes been charged upon the advocates of Inspiration, in consequence of their use of the term 'verbal,' as applied to this property of Scripture — a term which controversy has almost forced upon us in discussions on the subject. The origin of the phrase appears to be this: — Some have suggested that while the writers may have had ideas divinely infused into their minds, they were left altogether to their own unaided faculties and gifts to clothe them in appropriate language; and that, therefore, though the sublimest truths and revelations are to be expected in Holy Writ, yet human error and infirmity cannot be altogether absent. But, as if to meet such statements by anticipation, St. Paul expressly asserts that 'all Scripture (he speaks not here of the writers, but Scripture — that which is WRITTEN) is given by inspiration of God;' and many of the passages already quoted speak in the same strain. To express the opposite of the above-mentioned erroneous view, the term ' Verbal Inspiration ' has been brought into use; and, if

At the same time, we must not shut our eyes to the obvious scope and purpose of the communications retained, its origin should always be remembered. I understand it to imply simply that the language is inspired, as well as the truths, without at all entering into the question of the mode of this Inspiration. I conceive that, at least, the writers were so under the guidance and control of the Holy Spirit, as to preserve them from using language which would convey error.

Differences of style no objection.

It has been objected, that the existence of wide differences of style, so conspicuous in the writers of Scripture, militates strongly against the notion of the controlling influence implied in Inspiration. This controlling power, it is said, would produce an uniformity of style which is not found in Scripture. Those, however, who entertain such thoughts seem not to be aware that they are degrading the Divine Spirit to the level of one who possesses only the limited powers of a finite mortal. We beg, in reply, to ask such persons, What would you expect to be the style of the Holy Ghost? Are not all styles His? As He employs the free-will of men in all its tortuous varieties, to work out His purposes in the moral government of the world, cannot He overrule their natural powers and diversities of disposition and gifts, all in their freest action, for the holier and higher purpose of communicating His will? In this wondrous diversity we possess one of the most striking illustrations of the truth of that Inspiration for which we contend. The keys and stops may be as various as the minds of men; but when tuned by the Great Musician, and breathed upon by the Spirit from on high, each gives utterance to its own melodious strain, while all combine their richness to produce one harmonious whole. Accordingly, all the peculiarities of genius, character, thought, and feeling, belonging to the writers as human beings, and resulting from the several social positions they occupied, were employed by the Holy Spirit for communicating the Divine will in these records, written at various epochs through a long period of at least fifteen centuries from the days of Moses to the time of Christ, and yet breathing the same spirit, and manifesting a marvellous consistency under all their variety, in a way which a Divine original can alone account for.

Copyists not inspired.

It need scarcely be observed, that it is the Original Scriptures for which this claim of Inspiration is set up. We do not contend that

made to us in Holy Scripture, lest we should be looking for information which it was never intended we should

copyists were inspired, nor that they were necessarily free from those inaccuracies, in the process of transmitting the ancient manuscripts, to which the most careful are liable. We rejoice in the labours of the learned critic and philologist, who will, by comparison of MSS. and by researches into the meaning and use of words, improve the text, and bring it back as near to the original as possible. The result of their unwearied toil is a triumphant testimony to the almost perfect integrity of the text as we have it. Time, talent, and learning have been lavishly spent upon this laborious investigation; and the learned rationalist Eichhorn has admitted, that the different readings of the Hebrew MSS. collated by the indefatigable Kennicott (near 600 in number) offer no sufficient interest to compensate for the labour they cost.

Dr. Moses Stewart observes: 'In the Hebrew MSS. that have been examined, some 800,000 various readings actually occur as to the Hebrew consonants. How many as to the vowel-points and accents, no man knows. But at the same time it is equally true, that all these taken together do not change or materially affect any important point of doctrine, precept, or even history. A great proportion, indeed the mass of variations in Hebrew MSS. when minutely scanned, amount to nothing more than the difference in spelling a multitude of English words [*e.g.* קל for קול; as *honour*, for *honor*]. . . . Indeed, one may travel through the immense desert (so I can hardly help naming it) of Kennicott and De Rossi, and (if I may venture to speak in homely phrase) not find game enough to be worth the hunting. So completely is this chase given up by recent critics on the Hebrew Scriptures, that a reference to either of these famous collators of MSS., who once created a great sensation among philologers, is rarely to be found.'— Quoted in *Lee's Inspiration*, p. 410 (the most able, conclusive, and instructive work in the English language on the general subject of the Inspiration of Holy Scripture.)

Translators not Inspired.

Nor do we claim inspiration for the Translators. All sound researches, therefore, into the structure of language, and all discoveries in natural science, which help to throw light on the meaning of the original and to correct and improve our versions, are welcomed by the serious student.

It is for the Inspiration of the Original Scriptures, and for that

find; for it has been well said,* that the Scriptures are a 'record of the moral destinies of man, and therefore altogether unconnected with any exposition of the phenomena of the natural world, and the laws by which material things are held together.'

Some have conceived that in darker ages, ere the mind of man had unfolded itself to the varied wonders of the world around and within him which modern science has disclosed, a necessity existed for veiling truth in terms, not solely ambiguous and obscure, but imperfect, and even, in some instances, of doubtful verity, to meet the ignorance and prejudice of the times. But the necessity hinted at is purely one of their own invention. He who has all ideas, all language, all creation at His command—from whom all laws take their rule—to whom the past, the present, and the future, are all one, with all their occurrences—needs not to stoop to human imperfections in conveying His thoughts or the knowledge of His acts and works even to the most ignorant and illiterate. To the Infinite, the Omniscient, the Almighty, it is as easy to select terms which are in themselves correct, as words of inferior force—terms which will need no reforming hand to suit them to the endless changes which time unfolds, and which keener search and increasing light unceasingly add to the sum of human knowledge.

alone, that we contend; and the arguments we have adduced—which may be found drawn out at greater length, and with fuller illustrations, in works expressly on the subject—while they establish the fact for the Sacred Volume in general, at the same time involve the Inspiration of every portion of it.

* Sedgwick's *Discourse on the Studies of the University of Cambridge.*

Those, moreover, who feel inclined to regard this early portion of Scripture as parabolic, or given in vision, I would ask, What end is answered by such an hypothesis? That the Omniscient should afford us glimpses of futurity through the disguise of symbolic imagery and figurative language is what we might expect; for prophecies must be more or less obscure, otherwise they might be made a pretext for the most atrocious crimes. But the case is different with positive statements, assuming to be narratives of the past: if any statement at all were to be made respecting the past, there is no conceivable reason why it should not be the simple truth. The only effect of clothing facts in a fabulous disguise would be, to mystify what might just as well have been made perfectly intelligible.

What I contend for is, that the first eleven chapters of Genesis, being inspired by God, must be free from all error. Where terms are used and facts are affirmed, which belong even to the natural world alone, they can in no instance be wrong, nor involve any error; though they may communicate no philosophic truth and teach no physical law. We may have to modify our interpretations, to cast aside long-cherished prepossessions by which in our ignorance we and our forefathers had long enveloped and perverted the language of Scripture; but in the midst of all this (as I have attempted to show in the First Part by many examples), Holy Scripture still stands forth as the infallible Word of God, without blemish and without defect.

CHAPTER II.

ON THE SURPASSING IMPORTANCE OF THE FIRST ELEVEN CHAPTERS OF GENESIS.

It is possible that some who feel unable to gainsay the preceding argument for the Historical Character and Plenary Inspiration of this portion of the Word of God, and who are willing to admit that the discrepancies alleged against it are satisfactorily explained, may, nevertheless, have a lurking feeling that, after all, these earlier chapters are of comparatively small importance, and that Christianity would still stand intact were they even blotted out: and that, therefore, there is but little use in attending to them, or in taking such pains to vindicate them from the charge of being at variance in some of their statements with the facts of Science.

That this is a very mistaken view, I propose now to show. I have, in a previous page, pointed out in how many particulars our Lord and His Apostles have referred to these chapters, and drawn from them facts, arguments, and illustrations of the greatest importance. There are other matters to which I will now draw attention.

At the twelfth chapter of Genesis commences the history of Abraham and his descendants, which runs through the whole volume of the Old Testament to the end, relating the wonderful things God did in the preparation for the coming of the Saviour of the world. The first eleven chapters may be regarded, therefore, as the introduction to the rest of Scripture, the brief history of the world before the days of Abraham, and are a most precious relic of antiquity and treasure of divine revelation.

1. In the first place, as an HISTORICAL DOCUMENT this portion of Scripture stands unrivalled, as no other history in any nation under heaven can come near it in point of antiquity. The duration of the human race may be divided into three nearly equal portions of 2000 years each :— from Adam's creation to the time of Abraham is about 2000 years; from Abraham to the birth of Christ 2000 years; from Christ to the present day, nearly the same. As we ascend the scale backwards from the present day, we possess the history of some branches of the human race through the last of these periods up to Christ, and of Greece and Rome through one quarter only of the second or middle period; but it is Scripture alone which supplies any authentic or intelligible account of the long vista of years up to the beginning of this middle period; while in reference to the whole of the first period of 2000 years antiquity is profoundly silent: no vestige of history is to be found, except this sacred record to which I am now directing attention. We see, then, its great value in a merely historical point of view.

These chapters are of unrivalled antiquity.

2. But these chapters give us information concerning various most important matters: for example, the ORIGIN OF THE WORLD. How many conjectures and guesses on this subject do we find among the ancient philosophers of Greece and Rome, and also in the East? And yet how important it is to have some certain information regarding the relation of the Creator to the universe, that we may know our own position and our own duty.

<small>They tell us of the origin of the world.</small>

One class of Greek philosophers conjectured that matter is eternal; that all the order and harmony we see in nature is the result of chance; and that the gods take no concern in the affairs of the world.

The ancient Persians, before the change in their tenets brought in by Zoroaster, conceived that there were two independent first causes, the one light, or the good god, who was the author of all good, and the other darkness, or the evil god, who was the author of all evil, and that from the action of these two, in continual struggle with each other, all things were made.*

---

* 'According to the "Vendidâd" [the sacred book of the Parsees] Hormazd [the good deity] was opposed by Ahriman [the evil deity] in all his works. When Hormazd created Eriném véjo, similar to behisht, or paradise, Ahriman produced in the river the great adder, or winter; when he created Soghdo, abundant in flocks and men, Ahriman created flies, which spread mortality among the flocks; when he created Bakhdi, pure and brilliant in its colours, Ahriman created a multitude of ants, which destroyed its pavilions; when he created anything good, Ahriman was sure to create something evil. The power thus ascribed to Ahriman, that of creation, is greater than can be possessed by any created being, and the doctrine which teaches its exercise substantially promulgates the monstrous dogma of two eternal principles, which, though not unknown to the ancient Persians, is altogether unreasonable, as inconsistent with the predominance of order, regularity, and goodness

The Hindoos look upon the world as an *emanation* from the Deity himself, and, therefore, as part of the Deity. When Brahm, the Supreme Being, designed, they say, to produce the world, he threw off his abstraction, and became Bráhma, the Creator. From his mouth, arm, thigh, and foot, came the four castes, the priest, the warrior, the trader, and the labourer. As the fruit, they say, is in the seed, awaiting development and expansion, so all material forms exist in Bráhma: a notion which degrades the Creator to the level of the creature.

These are the opinions of men, feeling in their darkness after the truth, but lost in the dimness of unaided reason. But all these reasonings and guesses are set at rest by the revelation of God Himself in the opening of the Sacred Volume, and the empty speculations of philosophers may cease: 'IN THE BEGINNING GOD CREATED THE HEAVEN AND THE EARTH' (Gen. i. 1). There was, therefore, a BEGINNING, when nothing existed; and God CREATED OUT OF NOTHING all things that exist.

How sublime the idea! 'He commanded, and they were created' (Ps. cxlviii. 5). 'By the word of the Lord were the heavens made, and all the host of them by the breath of His mouth' (Ps. xxxiii. 6). It is thus by faith in God's declaration that we learn these things. 'By faith,' through the teaching of these chapters of Genesis, 'we understand that the worlds were framed

in the system of the universe, and altogether impious, as it leaves no being of infinite perfection, whom the mind of man may reverence, love, adore, and serve. The character ascribed to the works referred to, moreover, is totally inconsistent with their essential nature, &c.— *The Doctrine of Jehovah, addressed to the Parsees*, by the Rev. Dr. Wilson, Free Church of Scotland, 1839.

by the word of God, so that things WHICH ARE SEEN were not made of things WHICH DO APPEAR' (Heb. xi. 2), *i.e.* they were CREATED, not out of previously existing matter, but out of nothing.* This is an idea which no philosophy has ever conceived, and which Divine Revelation alone could teach us.

3. Take another question on which man's reason has puzzled itself in vain. How came EVIL INTO THE WORLD?

What Greek, or Hindoo, or other human system, ever gave an intellible account of this? But what Of the entrance of evil into the world. say these chapters? Here we have the narrative and the cause of this fatal event. How valuable, then, the record in which all is explained! God created Adam and Eve perfect: they were innocent and happy. But He made them not mere machines, to be His tools or instruments: while He created them with a holy bias on their minds, He gave them free wills. They were free to choose what was good, and to reject what was evil, or to do the opposite. As a test, there was one restraint laid upon them. They were not to eat of a particular tree; and God told them, that in the day they ate of it, they should die: it was 'the tree of knowledge of good and evil' (Gen. ii. 9); by eating of it they would disobey God, and would to their cost then know the good of

---

* I have seen the latter clause of this text quoted by itself, by an excellent scientific writer, in illustration of the idea, ' that invisible forces are behind and above all visible phenomena;' which is perfectly true, but hardly the meaning of this text. Is not its antithesis rather as follows, derived from the first clause?—' so that things which are seen were not made of things which do appear,' but out of nothing by the *spoken Word of God;*' ' He commanded, and they were created.'

obedience, and the evil of disobedience.* This one act of disobedience tainted their whole moral being. They fell, and evil entered the world.

We can understand this. If we are tempted to do some wrong action, and we resist the temptation and triumph over it, how we feel strengthened and encouraged! But suppose we yield, and commit the sin, is there not a sting within which afterwards goads us? Does not conscious guilt torment us? Has it not weakened our moral power? Conceive, then, the FIRST SIN, the sin of disobedience to an easy command of a good and gracious God. How, when once perpetrated, must it have stung the conscience of our first parents, and poisoned their moral nature! How forcibly and truly is this all set forth in the simple narrative of these earlier chapters of Genesis! How valuable, then, is this Record, in showing us, as no other can, how it came to pass that evil entered into the world!

4. There is another question. Although evil is come into the world, and has infected the whole human race, yet WHAT A WONDERFUL CREATURE IS MAN!

In his moral nature, how marvellous is the power of conscience, that inward monitor, accuser, and judge! Then, in his intellect, how great his genius, how varied his gifts, how wonderful his powers of expression! Observe the creations of his genius in music, in poetry, in painting, in sculpture; *Of the contrarieties in man.*

---

* 'Arbor itaque illa non erat mala, sed appellata est scientiæ dignoscendi bonum et malum, quia si post prohibitionem ex illa homo ederet, in illa erat præcepti futura transgressio, in qua homo per experimentum pœnæ disceret, quid interesset inter obedientiæ bonum et inobedientiæ malum.'—*Augustine de Genesi ad literam*, lib. VIII. cap. vi. § 12.

the triumphs of the reasoning faculty, by which, though tied to the earth, he can scale the heavens, and penetrate into the hidden laws which govern the universe; the marvellous gift of language, 'that miracle of human nature, at once its chief distinction and its highest glory,' and seen so remarkably in those renowned orators of ancient and modern times, who have moved the wills and passions of thousands by the power of eloquence; the instances of heroic self-devotedness which the history of the world furnishes; the many noble and lofty sentiments which philosophers have in all ages given utterance to—*and yet*, with all this, the MORAL DEGRADATION which marks man in every age; so that the mind shudders at the moral deformities which stain even the most polished times of antiquity, and at the opposition of principles which strive even in the Christian's breast, of whom it is justly said, that *when he would do good, evil is present with him* (Rom. vii. 21). Man is indeed a mystery of inconsistencies, a riddle of greatness and littleness, of good and evil. What account of this strange confusion of things has any philosophic system ever given? None. Pliny, in his perplexity, pronounced man to be an enigma not to be solved. Pascal notices these opposite principles thus: 'The greatness and misery of man being alike conspicuous, religion, in order to be true, must necessarily teach us that he has in himself some noble principles of greatness, and at the same time some profound source of misery.'

Turn, then, to the Holy Scriptures, and what do they tell us? Here we have the inspired history of facts which unlock the mystery, and show that man is

not in the state in which he was made, that he is a
RUIN—a monument of a once noble creature, bearing at
once the marks of his origin, and of the vast change
which has come over him. What do we read? First,
that 'God created man IN HIS OWN IMAGE, IN THE
IMAGE OF GOD created he him.' (Gen. i. 27.) He
created him in knowledge, righteousness, and true
holiness, after the divine resemblance; not infinite, but
pure, perfect, with high endowments, the glory of all
His creatures. But, secondly, that man by disobedience FELL from this image, and became a ruin. The
history of man's Origin, combined with the history of
his Fall, is essential to explain the condition and character of man as we find him now; and these are to be
found only in this sacred record.*

5. But let us cast back our thoughts once more
upon man as he came from his Maker's hands, 'the image and glory of God.' the crown of God's works, for whom creation *In showing man's origin they clear up other points.*
through former ages had been preparing, placed infi-

---

* 'A fall of some sort or other—the creation, as it were, of the non-absolute—is the fundamental postulate of the moral history of man. Without this hypothesis man is unintelligible—with it, every phenomenon is explicable. The mystery itself is too powerful for human insight. Such, in this matter, was the ultimate judgment of a man who in youth had entertained very opposite views—the poet Coleridge.'—Hugh Miller's *Testimony of the Rocks*, p. 265.

'It is astonishing that the mystery which is farthest removed from our knowledge (I mean the transmission of original sin) should be that, without which we can have no knowledge of ourselves. It is in this abyss that the clue to our condition takes its turns and windings, insomuch that man is more incomprehensible without this mystery, than this mystery is incomprehensible to man.'—Pascal quoted in Dr. McCosh's *Method of the Divine Government*.

'A paradise, a condition of primeval innocence, a state of proba-

nitely above the world by being thus made in the divine likeness.

No philosophy has ever so well explained the proper basis of physical Science, as the Scriptures here do; which is, that as God has made the world by His Word or Wisdom, so man, being made in the image of God, is created with faculties which put it in his power to explore and to systematize those works. Here we have an answer at once to all sceptical questions about the reality of sensation and of science.

See, too, the vastness of man's moral capacities as thus created, and the love bestowed upon him at his creation, which is the basis of the whole doctrine of Redemption. This does away with the supposed incredibility of an Incarnation; for the end was none other than that which had been the ultimate design of all previous dispensations—a restoration of the Image of God in human nature.

6. Next see in this simple narrative of Scripture how with a master-touch the very essence of successful temptation to sin is exposed to view, and the source of all that disobedience to God which so mars the moral aspect of the world.

They detect the essence of successful Temptation

tion, and a *fall*, are absolutely requisite before we can explain anything connected with man. Without these, philosophy would lead us only to a hopeless mystery; we should know absolutely nothing, and never should be able to attain to knowledge; for all the science that has ever been evolved does not advance man a single step in the explanation of his moral nature and moral condition. No man who has rejected these four particulars has ever been able to advance an explanation possessing even the most remote claims to acceptance. They, and they only, solve the perplexing question of human existence—of man endowed with the conception of the virtues, yet constantly practising the opposite evils.'
—Dove's *Logic of the Christian Faith*, p. 352.

What was the snare which Satan so successfully laid to entrap our first parents? What is the bait which so easily draws us from our allegiance to God who made us, and who preserves and blesses us? Is it not summed up in these words—'YE SHALL BE AS GODS?' 'When the woman saw that the tree was . . . to be desired to MAKE ONE WISE, she took of the fruit thereof, and did eat;' although God's command was express, 'Ye shall not eat of it.' (Gen. iii. 3, 6.) Here was the assailable point, the fancied glory of INDEPENDENCE, the heart's revolt against restraint, an uncreature-like rebellion of the will against dependence even upon God, whose service is perfect freedom.

How much is this at the root of all our rebellion now! It was the deification of man's reason, the atheistic pride of vice and intellect combined, the blasphemous assumption of an independence of all that was holy, divine, and true, that distinguished the Infidels of the French Revolution of 1798, as well as of Voltaire and his accomplices, who came to such a wretched end. It was the pride of independence which hurled the devil and his angels down to hell. And how wonderfully does this record of Holy Scripture put it prominently forward as the master-stroke, with which the arch-deceiver plied his art with such awful success against our first parents—innocent, but not invulnerable!

7. These chapters contain some remarkable points in History. Here we find the first institution of Marriage, and of the Sabbath. They give the only rational account of that great event, the Deluge, the reflex of which is seen in the *They contain remarkable points in History.*

innumerable traditions among nations the widest apart and the most dissimilar in habits and character. They explain, by a simple history of the divine interposition in the dispersion of mankind, because of their pride and wickedness, the singular results at which philologists have of late arrived regarding the 6,000 languages and dialects at present spoken—that such is the internal relationship of their radical words and inflections and constructions, that there is every reason for supposing that they must have proceeded from one primitive tongue, and that the separation into branches must have arisen from some violent and sudden cause—a theory which is a remarkable comment upon the history of the Tower of Babel.

This sacred record gives the only history of the apportioning the earth to the several nations, as we now find them. After an enumeration, 'These,' says the inspired writer, 'are the families of the sons of Noah, after their generations, in their nations; and by these were the nations divided in the earth after the flood.' (Gen. x. 12). To which St. Paul, 2000 years afterwards, alludes, 'God hath made of one blood all nations of men for to dwell on all the face of the earth, and hath determined the times before appointed, AND THE BOUNDS OF THEIR HABITATION.' (Acts, xvii. 26.)

Here, too, we have a certain clue to the first institution of SACRIFICES — that remarkable method for attempting to appease the Divine wrath, which we find prevailing all the world over, wherever Christianity has not shed its light and superseded the type by substituting the sacrifice of the Redeemer Himself.

8. Lastly, in these wonderful chapters we have the germ, the rich and fruitful germ, of all Prophecy, in two of the most remarkable and comprehensive predictions which the whole Scripture contains :— *They contain the germ of all Prophecy.*

One, the promise of the Seed of the Woman, who should bruise the serpent's head while he should bruise his heel—a prophecy which has been so largely fulfilled in the miraculous birth of Christ of a pure virgin, who by His death upon the cross, His triumphant resurrection and ascension to the Father's throne, has overcome sin and death, and opened the kingdom of heaven to all believers—a prophecy which is fraught with blessings to all nations, and which will be completed in its fulfilment when He comes the second time to judge the world, and to reign for ever and ever, King of kings and Lord of lords :—

The other Prophecy—the curse upon Canaan, and the blessing upon Shem and Japhet, spoken in few words, but carrying with them, as years after years still roll on, their own evidence of divine inspiration by the wonderful accomplishment they are perpetually receiving.*

Who can deny, then, that this portion of Holy Scripture is a treasury of unspeakable value, and worthy of its high origin? The simplicity of the narratives, combined with the surpassing importance of the truths conveyed in them, is a confirmation of their authority. Who but one writing under the

* For an able and very instructive article on the accomplishment of this prophecy, see *Journal of Sacred Literature*, No. XXVI., July 1861.

guidance of inspiration could, not merely have *avoided* all the inanities of human cosmogonies, but have delivered at that early period, and in the most unassuming way, accounts of the most momentous transactions, which are found to harmonize with the most recent investigations of science, and which supply the most profound information on theological and moral subjects?

CONCLUSION.

WHAT, then, are the results arrived at in the foregoing pages? They may be summed up under the following heads:—

1. That, through ignorance and hasty zeal, Holy Scripture has undergone many severe tests during the progress of Science, and has come through the trial in every case with triumph. The experience of the past has worked out this result, that through the whole course of philosophical discovery, Scripture and Science have never been found at variance, though they have often been charged with being so.

2. That Scripture speaks in human language, and according to its usages; but in no case adopts the errors and prejudices of men, even in things natural. It speaks to us on such matters according to the appearances of things, that is, as things ARE SEEN, which is a way intelligible in all ages of the world. It speaks as man would speak to man in every-day life, even on such topics, and in times of the greatest scientific light. It speaks not scientifically, and therefore does not adopt scientific terms, or give scientific views of things: but there is, nevertheless, no sacrifice even of scientific truth to human ignorance and prejudice.

3. That this harmony between Scripture and Science appears, not only from the abundant illustration it receives from the history of past conflicts through which this Sacred Volume has passed intact, but pre-eminently from the character of Scripture itself as the Inspired Word of God, and therefore infallible in every respect.

4. That the Earlier Chapters of the Sacred Volume, in which the seeds of variance have been supposed to lie, are of inestimable value to us; and the fact of their Inspiration must not be set aside on the pretence that Christianity would remain the same if they were blotted out; for they form a most important portion of the Divine Revelation, and convey inspired truths of the highest moment.

The Conclusion, then, which I would draw in these days of advancing knowledge is this, THAT NO NEW DISCOVERIES, HOWEVER STARTLING, NEED DISTURB OUR BELIEF IN THE PLENARY INSPIRATION OF SCRIPTURE, OR DAMP OUR ZEAL IN THE PURSUIT OF SCIENCE. That difficulties should have existed, and should still occasionally appear, is not to be wondered at when we consider the things which are brought into comparison. On the other hand, here is an ancient record handed down to us, first by the careful guardianship of the Jewish nation, and now in addition by that of the Christian Church, referring to events which occurred 6000 years ago—intended for the instruction of every successive age, from the period when it was written down to the end of time. The authority of this record is attested by our Lord and His Apostles, inasmuch as in nearly seventy passages they quote or refer to it as authoritative. The great truths taught in it, are such as the acutest philosophers of Greece and Rome and the East failed to discover—such as the origin of the world ; the origin of man ; how evil came into the world ; the promise of a Saviour ;—and with these revealed truths are mixed up statements regarding the aspect and condition of the earth and of the heavenly bodies around. The language of this record was fixed thousands of years ago, and is brief and terse, and not designed to communicate to us scientific information, but only those great realities which reason could never have found out. On the other hand, Science has been opening the book of Nature, and turning over its countless leaves with unwonted rapidity. Theories have been built up, imperfect in their parts, and too often with hasty generalizations. And a harsh and crude comparison is made with this venerable document, which speaks not the language of a varying and growing Science, but is couched in terms to be intelligible in all ages. Is it to be wondered at, that as new and ill-digested physical facts rise upon our view with such rapidity, frequently combined by gratuitous hypotheses, contradictions between Scripture and Science should be perpetually appearing? It is clear to what cause this is to be attributed—not to Scripture, but to a restless and imperfect Science. The experience of the past teaches us this. The hasty and immature deductions of Science may sometimes stand in opposition to Scripture : but those settled results in which the body of philosophers agree, often confirm and illustrate the statements of the Inspired Volume.

Let us, then, hold firm our grasp upon this truth, that the Scriptures are the infallible Word of God, true in every statement they contain, although the interpretation sometimes demands more knowledge than we at present possess; but let us at the same time remember, that there is no ground whatever for ceasing to pursue Science, in all its branches, with an ardent and fearless mind. God's Word and Works never have contradicted each other, and never can do so. Some would decry the pursuit of Science as endangering revelation; they tremble for the result, as new discoveries are announced and reason publishes its triumphs. But these are short-sighted and ill-placed apprehensions: nor would such a course remedy the evils feared. The progress of Science is inevitable. As well might we desire to hold back the wheels of time, or attempt to enchain the thoughts of men, as to arrest its course. The progress of Science is indeed the glory of man's intellectual endowments; and to live in ignorance of the history and material laws of the universe, of which he forms a part, is a libel on that commanding gift with which God had endowed him, rendering him pre-eminent above the rest of His creatures. The progress of Science is the setting forth of the greatness and wisdom of the Creator in His works: and to desire to check it, or to fear its results, is to betray our narrow prejudices, and to refuse to recognise the hand of GOD in His own world. Let us, therefore, push our investigations to the utmost with untiring energy.

Let us not shrink, moreover, from stating our difficulties in their broadest features, and laying open without hesitation all that appears contradictory. We have nothing to fear. The greatest perplexities may at any time surround us; but both reason and experience have armed us with arguments, which assure us that all will be right. Whatever happens, let our persuasion always be avowed, that Scripture cannot err. Let us be content rather to remain puzzled, than to abandon, or even question, a truth which stands upon so immovable a basis.

It is the doubts and surmises of those who are looked to as authorities in these matters, which shake men's minds. It is the hazardous assertions* of some

---

* Take, for example, such statements as the following: 'In a former Essay I have adverted to the question of discrepancies between Science and the language of Scripture generally, and have referred more especially to that notable instance of it—the irreconcilable contradiction between the whole view opened to us by Geology and the narrative of the Creation in the Hebrew Scriptures, whether as briefly delivered from Sinai, or as expanded in Genesis. In the minds of all *competently informed persons* at the present day, after a long struggle for existence, the literal belief in the Judaical cosmogony, it may now be said, has died a natural death' [!]—And after some remarks on a theory of evolution or progressive development of animal life, he adds :—
'Those who accept geological truths at all, and admit their palpable contradiction to the Old Testament, without prejudice to their faith, cannot with *consistency* make it a ground of objection to any hypothesis of the *nature* of the changes indicated. that they are *contrary to Scripture*. They are in no way *more* so than *all geology* is' [!]—*Essay on the Philosophy of Creation.* By the Rev. Baden Powell, Savilian Professor of Geometry in the University of Oxford.

My answer to all this is:—There is in the sense intended, no 'narrative of the creation in the Hebrew Scriptures,' no 'Judaical Cosmogony,' no 'contradiction of geological truths to the Old Testament,' according to the principles of the foregoing Treatise.

In a subsequent publication, *The Order of Nature*, Professor Powell made some remarks upon the present Treatise (second edition); regarding which I will here state that, through some unaccountable misconception, he has made me say the reverse of what I do say. His words are :—' The whole view of the subject presented in the work referred to is this :—As the Scripture in former times *seemed* opposed to the

who occupy the seat of the philosopher—who ought to be models* of philosophical prudence, holding even the scales of truth amidst the headstrong and uninformed—which create the confusion.

Such a course is UNPHILOSOPHICAL in the highest

motion of the earth and the existence of antipodes . . . so, again, it *seems* to teach a common origin of the human race—a common primeval language—and other similar tenets. But these, we are told, are only "*false conclusions* deduced by its votaries ;" false *interpretations*, which yet are identical with the *very words;* fallacious *conclusions*, which, notwithstanding, are directly asserted in the very terms ; and nevertheless, the historical authority of the passages, and their surpassing importance, are to be strictly maintained ' (pp. 223, 224).

What I do say may be seen under the Third Class of Examples in Part I., at p. 104 and the following pages of this edition, which differ in no respect from the second edition but in the addition of new matter. It will there be seen, that I state that Scripture *does* teach, not that it *seems* to teach, the common origin of the human race, and other similar tenets. And my readers ' are told ' that Science (*not Scripture*), through the false teachings of some of its votaries, has been made to say the opposite to this, and therefore has been forced into apparent collision with Scripture ; but that wiser men have delivered Science from this reproach by showing that these conclusions of its votaries are false ; and that, in fact, Science does not, in these or any other cases, lead to results opposed to Holy Scripture. This is the reverse of what Professor Powell has made me say. This ' outrageous ' treatment of my words was noticed at the time in the *Quarterly Review*, October 1859, p. 431, note.

* See some excellent remarks, written in this spirit, in the chapter on the relation of Tradition to Palætiology, *in* Dr. Whewell's *Philosophy of the Inductive Sciences*. Also the following sensible remarks from the lips of another philosopher, Professor Stokes, when President of the British Association in 1869 :—

'Truth, we know, must be self-consistent, nor can one truth contradict another, even though the two may have been arrived at by totally different processes, in the one case, suppose, obtained by sound scientific investigation, in the other case taken on trust from duly authenticated witnesses. Misinterpretations of course there may be on the one side or on the other, causing *apparent* contradictions. Every mathematician knows that in his private work he will occasionally by

degree; and not only so, it is MISCHIEVOUS in every way. It unsettles the minds of the young; it plays into the unbelieving prepossessions of the infidel; it confirms the sceptic in his disregard of religion. On the other hand, it repels the timid from the pursuit of Science; it disunites, instead of harmonizing; it checks the progress of truth; it sets at variance things which in reality agree.

two different trains of reasoning arrive at discordant conclusions. He is at once aware that there must be a slip somewhere, and sets himself to detect and correct it. When conclusions rest on probable evidence, the reconciling of apparent contradictions is not so simple and certain. It requires the exercise of a calm, unbiassed judgment, capable of looking at both sides of the question; and oftentimes we have long to suspend our decision, and seek for further evidence. None need fear the effect of scientific inquiry carried on in an honest, truth-loving, humble spirit, which makes us no less ready frankly to avow our ignorance of what we cannot explain than to accept conclusions based on sound evidence. The slow but sure path of induction is open to us. Let us frame hypotheses if we will: most useful are they when kept in their proper place, as stimulating inquiry. Let us seek to confront them with observation and experiment, thereby confirming them or upsetting them as the result may prove; but let us beware of placing them prematurely in the rank of ascertained truths, and building further conclusions on them as if they were.'

INDEX.

Abbeville, 190.
Abd-al-latif, 179.
Acheul, St. 190.
Agassiz, Professor, 124, 187, 204.
Age of human race, 151, 155, 176, 180.
Age of Vedas, 153.
Ages of stone, brass, and iron, 183, 199.
Amiens, 190.
Angels, agency of, 281.
Anglo-Americans, 123.
Animals and plants long before Adam, 42.
Antipodes, 18.
Antiquity of the earth, 42.
Appearances, meaning of, 29.
Aquosity and vitality, 205.
Archimedes, 27.
Architecture of India, 198.
Argyll, Duke of, 81, 220, 240, 250-253, 283, 293.
Aristotle, 24, 26.
Arithmetical objections, 306.
Astronomy, Chinese, 154, 328.
    ,, Greek, 153.
    ,, Hindu, 151, 328.
Augustine, 18, 21, 25, 361.
Auriguac, 191.
Auvergne, 203.
Babylon, 166, 327.
Baillie, M. 151.
Bara and Asah, 47.
Bashan, 166.
Belshazzar, 166, 327.
Bentley, Mr. 150.
Berosus, 150, 166.
Biot, M. 155.
Birks, Rev. T. R. 56, 316.
Birs Nimroud, 151.
Bize, 188.
Bockh, Professor, 168.
Bopp, Professor, 145.
Borsippa, 151, 166.
Boucher de Perthes, M. 181, 190.

Brain capacity, 237.
Brewster, Sir D. 27.
British Association for the Advancement of Science, 94, 255, 264, 276.
Brixham, 190.
Bronze age, 183, 197.
Buckland, Rev. Dr. 45, 95, 230.
Buckle, Mr. 276.
Buffon, M. 298.
Bunsen, M. 156.
Bushmen, 253.
Cagliari, 191, 206.
Caldwell, Rev. Dr. 144, 146.
Calvin, 24.
Canoes in Scotland, 187, 205.
Carpenter, Dr. 110, 219.
Cashmere, 187.
Caucasian type, 112.
Challis, Rev. Professor, 65.
Chalmers, Rev. Dr. 45, 50, 284.
Changes in type, 115.
    ,, *per saltum*, 140.
Chili, 187.
Chimpanzee, 302.
Chinese Astronomy, 154, 328.
    ,, Chronology, 155.
Colenso, Dr. 35, 48, 80, 92, 306, 334, 335, 338.
Compte, M. 298.
Confusion of tongues, 142.
Confucius, 154.
Contrarieties in man, 361.
Copernicus, 32.
Corn needs cultivation, 92.
Cornwall, 188.
Cosmos, monk, 20.
Cuvier, M. 296, 297.
Danish peat and shell mounds, 183.
Darwin, Mr. 221, 241-249, 254, 295, 296.
Dathe, 45.
Davison, Rev. Mr. 101.
Dawson, Dr. 67.

Days of Creation, two ways of explaining, 44.
Death before Adam, 86.
Delitzch, Professor, 92.
Deluge, 94, 96.
Denise fossil man, 191.
Deshayes, M. 83, 217.
Design, 293.
Differences between man and apes, 236.
Difficulties and obscurities exaggerated, 101.
D'Orbigny, M. 60, 75, 83, 217.
Dove, Dr. 364.
Dowler, Dr. 186.
Dravidian languages, 146.
Druids, 205.
Earth, Antiquity of, 42.
„ Form of, 21.
„ Motion of, 23.
Egyptian Antiquities, 155, 328.
„ Paintings, 138.
Egyptology, 155, 164, 172, 176.
Eichhorn, 353.
Ellicott, Bishop, 88.
Elliptic type of crania, 112.
Engis fossil skull, 189.
Eras past and present compared, 323
Eratosthenes, 163.
Esquimaux, 253.
Evil, entrance of, 360.
Evolution, 221, 234, 245, 256, 262.
Falconer, Dr. 190.
Final causes, 293.
Finns, 117.
Firmament, 14.
Flint remains of man, 180, 190, 253.
Florida reefs, 187.
Form of the earth, 21.
Fowler, Rev. F. W. 316.
Frere, Mr. 191.
Galen, 299.
Galileo, 23.
Gaudry, M. 261.
Gaussen, M. 28.
Genesis, chap. i. not a vision, 55.
„ Chap. x. 132, 145.
„ Earlier chapters of, 329.
„ First chapter, 294.
„ First and second agree, 92.
„ First three, 90.
„ First eleven, 333.
„ Historical character and inspiration of, 335.
„ Surpassing importance of, 356.
Genius and inspiration, 350.

Geological record imperfect, 226, 259, 260.
Geology, 41, 84.
Goodwin, Mr. 17, 29, 41, 54, 78.
Gorilla, 237, 302.
Gower, 191.
Graham, Mr. Cyril, 168.
Granite, 204, 219.
Gravitation, Law of, 263, 271, 291.
„ and development compared, 263.
Greek astronomy, 153.
Grote, Mr. G. 165.
Grove, Mr. 255.
Hauran, 166.
Hebrew chronology, 158.
Hesiod's ages, 199.
Hieroglyphics, 169.
Hincks, Dr. 169.
Hindu astronomy, 151, 328.
„ notion of Deity, 359.
Hitchcock, Professor, 87.
Hodgson, Mr. 134.
Hopkins, Mr. W. 264.
Horn, Cape, 196.
Horner, Mr. L. 176.
Hottentot race, 121.
Hoxne, 191.
Humphreys, Colonel, 222.
Hungarians, 116.
Huxley, Professor, 63, 76, 85, 110, 136, 140, 238, 254, 257, 259, 260, 262, 263, 264, 266, 270, 295, 301.
Icklingham, 191.
Illogical geology, 84.
Imperfect theories, Danger of, 82.
Indo-European languages, 144.
Inspiration defined, 335.
„ of Old and New Testaments, 348.
Irish Lake-dwellings, 185.
Iron age, 183.
Josephus' chronology, 159.
Joshua's miracle, 35, 338.
Kant, 63.
Keil, Professor, 92.
Kennicott, Dr. 353.
Kepler, 27.
King, Professor W. 205.
Kirkdale cavern, 205.
Lactantius, 22.
Lamark, M. 220, 232.
Languages, Families of, 144.
„ Origin of, 148.
„ Unity of, 142.
La Peyrère, M. 127.
Laplace, 62, 298.

INDEX.

Lapps, 117.
Lardner, Dr. 75.
Latham, Dr. 118, 134.
Law, On supremacy of, 276.
Lee, Archdeacon, 353.
Lecky, Mr. 21, 127, 284.
Lewis, Sir G. C. 27, 164, 176, 329.
Liege, 189.
Light before the six-days, 43.
,, Undulatory Theory of, 271, 292.
Livy, 326.
London-wall excavations, 213.
Lubbock, Sir J. 198, 213, 235, 251.
Lunar mansions, 153.
Lyell, Sir C. 83, 85, 182, 192, 202, 210, 217, 219, 220, 232.
McCaul, Rev. Dr. 17, 47, 62, 92.
McCausland, Dr. 68, 73.
McCosh, Dr. 74, 363.
Macmillan, Rev. Hugh, 94.
Magyar race, 116.
Malacca, Straits of, 184.
Man, Antiquity of, 151, 155, 176, 180, 234.
,, and apes compared, 236.
,, Contrarieties in, 361.
,, First, not a barbarian, 250.
,, Has he several origins? 124.
,, Origin of, 234.
Manetho, 163.
Mansel, Dean, 292.
Malayo-Polynesian races, 122.
Matter and mind, 270, 271.
Mayhew, Mr. H. 122.
Memphis, 176.
Miller, Mr. Hugh, 53, 61, 74, 97, 363.
Milton, 28, 326.
Miracles of Joshua, 35, 338.
Miracles not incredible, 286.
Mississippi delta, 180, 202.
Moabite stone, 166, 328.
Moore, Dr. G. 111, 141, 233, 236.
Morlot, M. 185.
Müller, Mr. Max, 148.
Murchison, Sir R. 226.
Naples, 187.
Natchez, 192.
Natural-day theory, 44, 74.
Natural selection, 221, 223, 294.
Nature, Uniformity of, 275.
,, Use of the term, 305.
Neanderthal human fossil, 189.
Nebular hypothesis, 63.
Negro, 118.
Nepaulese, 131.
New-Englanders, 123.

New school of historical criticism, 324.
Newton, Sir Isaac, 232, 234, 262, 324.
Niagara, 203.
Niebuhr, 324, 326.
Nile deposits, 176, 185, 328.
Nineveh, 166, 327.
Nomadic tribes, 122.
Norris, Mr. 118.
North Atlantic deep-sea dredgings, 219.
Norway, 188.
Nott, Dr. 124.
Oceana, 121.
Ohio, Valley of, 185.
Oise, 190.
Ollivant, Bishop, 213.
Oppert, M. 150.
Origin of life, 254.
,, Man, 104, 234.
,, Species, 221.
,, the world, 358.
Osburn, Mr. 179.
Ouse, 190.
Owen, Professor, 215, 299.
Paley, Archdeacon, 299.
Pascal, 363.
Patagonians, 253.
Pattison, Mr. 193, 210, 211.
Peat in Denmark, 183.
,, Growth of, 193.
Pentateuch, 306.
Period-day theory, 52.
,, Objections to, 56.
Persians, Ancient, notion of Deity, 358.
Peru, 187.
Phenomena, Method of describing, 29.
Philolaus, 27.
Physical world not abandoned to physical laws, 278.
Pitcairn Islanders, 136.
Pliny, 205.
Polynesian races, 121.
Pondres, 188.
Pool, Matthew, 100.
Positivists, 297.
Powell, Rev. Professor B. 373.
Prayer and order of nature, 280–282.
Pre-adamite species, 42.
Prestwich, Mr. 190, 208.
Pritchard, Dr. 104, 138.
Prognathous type, 111.
Protoplasm, 257, 265–269.
Ptolemaic system, 32.
Ptolemy, 26.
Pusey, Rev. Dr. 79, 89.

Pye-Smith, Rev. Dr. 217.
Pyramidal type, 111.
Pythagoras, 27.

Quotations from, or References to Books:—
Adam and the Adamite, 127.
Aids to Faith, 47, 62, 146, 293.
Ancient Egyptians, 179.
Antiquity of Man, 84, 140, 182, 218, 232, 237.
„ Examination of, 194, 210.
Archaia, 68.
Astronomy of the Ancients, 27, 164, 165, 176.
Bampton Lectures, Rawlinson, 146.
Bible Teachings in Nature, 94.
Blackwood's Magazine, 210.
Bridgewater Treatises, 45, 230.
British Quarterly Review, 135.
Cambridge Essays, 168.
Cesar's Commentaries, 195.
Chips from a German Workshop, 47.
Christian Observer, 18, 28, 60, 205.
„ Remembrancer, 89.
Colenso on Pentateuch, 306.
Comparative Grammar of Dravidian Languages, 144, 146.
„ Indo-European Languages, 145.
Considerations on the Pentateuch, 300, 318.
Creation in Plan and Progress, 66.
Contemporary Review, 267.
Credibility of Early Roman History, 320.
Cyclopedia of Anatomy and Physiology, 76, 123.
Darwinian Theory Examined, 295.
Descent of Man, 241.
Destiny of the Creature, 89.
Doctrine of Jehovah, 359.
Dublin University Magazine, 169.
Early History of Mankind, 252.
Egypt's Place in History, 156, 170, 174.
Encyclopedia Britannica, 155.
Essays and Reviews, 17, 29, 276, 277, 280.
Exodus of Israel, 310.
Fertilisation of Orchids, 290.
First Man, and his Place in Creation, 111, 141, 233, 236.
Fortnightly Review, 136.
Fraser's Magazine, 264.
Gaussen's Theopneustia, 28.

Quotations, &c. continued:—
Geological Quarterly Journal, 180.
Gesenius' Hebrew Lexicon, 16, 69.
Goodwin's Mosaic Cosmogony, 29, 41, 54.
Grant's Astronomy, 63.
Grote's History of Greece, 165.
Guardian, 323, 336.
Hale's Chronology, 162.
Herschel's Astronomy, 63.
Hindu Astronomy, 152.
History of Civilization, 256.
History of Inductive Sciences, 22, 40, 296.
History of Rome, Niebuhr, 324, 326.
Humboldt's Cosmos, 63.
Inspiration of the Bible, 337.
Josephus' Antiquities, 15.
Journal of Sacred Literature, 367.
Journal of Victoria Institute, 101.
Lay Sermons, Addresses, and Reviews, 63, 76, 110, 140, 222, 264, 266, 267, 295.
Lee's Inspiration, 353.
Letter on Dr. Colenso's Part I. 313.
Literary Churchman, 17.
Logic of the Christian Faith, 364.
London Labour and London Poor, 122.
Macmillan's Magazine, 199, 205.
Man's Place in Nature, 238, 301.
Martyrs of Science, 27.
Medical Times and Gazette, 213.
Museum of Science and Art, 60, 75, 218.
Natural History of Man, 104.
New Facts and Old Records, 211.
Order of Nature, 373.
Origin of Species, 221, 226, 229, 254, 264.
Philosophical Magazine, 165.
Philosophical Transactions, 177, 193, 209, 222, 300.
Philosophy of Creation, 373.
Philosophy of Inductive Sciences, 40, 374.
Physical Basis of Life, 270.
Pool's Synopsis, 16.
Popular Science Review, 77.
Pre-adamite Man, 127.
Pre-historic Times, 198, 213, 235.
Primeval Man, 81, 226, 240, 250–252.
Primitive Condition of Man, 235.
Principles of Geology, 83, 182, 194, 220, 232.

INDEX.   381

Quotations, &c. continued:—
Proceedings of Royal Society, 220.
Prodrome de Paleontologie, 60, 75.
Pusey's Daniel, 79, 80.
Quarterly Review, 164, 198, 204, 374.
Rationalism in Europe, 21, 127.
Record, 202.
Recovery of Jerusalem, 108.
Reign of Law, 283, 293, 299.
Replies to Essays and Reviews, 56.
Saturday Review, 199.
Science of Languages, 148.
Sermons in Stones, 69.
Shinar, 129.
Smith's Dictionary of the Bible, 80, 151.
Supernatural in relation to Natural, 74.
Surya Siddhanta, 152, 154.
Testimony of the Rocks, 53, 60, 75, 95, 97, 99, 124.
Types of Mankind, 130, 230.
Unity of the Human Races, 124.
Universal Review, 84.
Varieties of Man, 111.
Wallace on Natural Selection, 112, 139, 224.
Week of Creation, 71.
Wordsworth's Genesis, 80.

Raamses, 313.
Races, adapted to climates? 133.
 „ Can all mix? 136.
 „ Divergence since the Flood, 138.
Raleigh, Sir W. 98.
Rameses, 176, 313.
Rawlinson, Professor, 146.
 „ Sir H. 166.
Reconciliations, Forced, not advisable, 101.
Redi, Francesco, 264.
Reproduction indicates vitality, 267.
Resurrection, Objection to, 319.
Rigillot, Dr. 190.
Robillas, 131.
Roman History, Early, 326.
Rorison, Dr. 56.
Rosetta Stone, 169.
Rütimeyer, Professor, 261.
Saint-Hilaire, M. Geoffroy, 295.
Samaritan Pentateuch Chronology, 159.
Santos in Brazil, 186.
Schmerling, Dr. 189.

Science confirms miracles, 292.
 „ Prospects of, overrated, 289.
Scripture allusions to natural phenomena, 102.
 „ Chronology, 157.
 „ Phraseology, Wisdom of, 27, 36.
Sedgwick, Rev. Professor, 45, 354.
Seine, 190.
Semitic languages, 144.
 „ nations, 199.
Septuagint chronology, 159.
Shishak, king of Egypt, 173, 327.
Silvestre de Sacy, 180.
Smyth, Dr. T. 124.
Somme, 190.
Sothiac cycle, 167.
Species, Extinction of, 208.
 „ Mutability of, 225.
 „ Of human period, 216.
 „ Origin of, 221.
Specific centres, 88.
Stanley, Dean, 314.
Stewart, Dr. Moses, 353.
Stillingfleet, Bishop, 100.
Stokes, Professor, 374.
Stone age, 183, 195.
Strata, Comparative thickness of, 211.
Sun-rise and Sun-set, 31.
Sweden, 188.
Swiss Lake-dwellings, 184.
Syncellus, 163.
Tamil, 147.
Taylor, Bishop Jeremy, 87.
 „ Mr. Isaac, 300, 318.
 „ Mr. John, 205.
Teloogoo, 147.
Temple, Bishop, 276, 289.
Temptation, essence of, 364.

Texts quoted or illustrated:—
Genesis, i. 1. 26, 48, 52, 80, 234, 359.
 „ 2. 74, 77, 80, 330.
 „ 3. 44, 339.
 „ 5. 52, 56, 60.
 „ 6-8. 16.
 „ 7. 16.
 „ 9, 10. 92, 339.
 „ 11. 61, 70, 339.
 „ 11-13. 53.
 „ 11-27. 227, 255.
 „ 14, 15, 17. 16, 294.
 „ 16. 57.
 „ 16-18. 44.
 „ 18-24. 294.
 „ 20. 69, 93.

Texts quoted, &c. continued:—
Genesis, i. 20–23. 52.
„ 20–28. 92.
„ 21. 17, 70.
„ 24, 25. 52.
„ 25. 93, 294.
„ 26. 68, 339.
„ 26, 27. 48.
„ 26–28. 94.
„ 27. 92, 93, 94, 340.
„ 28. 94.
„ 28, 29. 58.
„ 29. 61.
„ 29, 30. 294.
„ 31. 60, 294.
„ ii. 1–3. 340.
„ 4. 52, 90.
„ 4, 5. 66.
„ 5. 21.
„ 6. 92.
„ 7, 8, 10. 47, 92, 93, 234, 274, 340.
„ 9. 340, 366.
„ 10. 92.
„ 15. 94.
„ 16. 46.
„ 18. 341.
„ 19, 20. 67, 93.
„ 23, 24. 341.
„ iii. 4–6. 341, 365.
„ 15–20. 342.
„ 16. 274.
„ 20. 104.
„ iv. 4, 8. 342.
„ 10. 343.
„ 16, 17. 126.
„ 25. 126.
„ v. 3. 126, 343.
„ 3–28. 158, 353.
„ 4. 127.
„ 20–28. 255.
„ vi. 3, 14, 17, 18. 343.
„ 5–8. 157.
„ vii. 11. 16.
„ 11, 12. 97.
„ 19. 100.
„ viii 2, 3. 98.
„ 13. 158.
„ 22. 278.
„ ix. 3, 4. 343.
„ 11. 344.
„ x. 2. 94.
„ 6. 172.
„ 12. 366.
„ 15. 157.
„ 32. 104, 344.
„ xi. 1. 150.

Texts quoted, &c. continued:—
Genesis, xi. 10–26. 158, 159, 160, 344.
„ xii. 4. 159.
„ 10–20. 172.
„ xv. 13, 16. 159, 160, 317.
„ 18. 315.
„ xvi. 7. 282.
„ xix. 1–25. 282.
„ xxxii. 1, 2. 282.
„ xxxix. 1. 172.
„ xlvi. —. 309, 317.
„ 20, 27. 309.
„ xlvii. 11. 313.
„ 28. 310.
Exodus, i. 11. 313.
„ 15. 173.
„ vii. 7. 318.
„ 9, 10, 12. 18.
„ viii. 3. 69.
„ xi. 4. 314.
„ xii. 3, 6, 39. 313.
„ 12. 314.
„ 36. 312.
„ 37. 313.
„ 40. 159.
„ xiii. 1, 2. 315.
„ 2, 12. 316.
„ 18. 312.
„ xix. 1. 161.
„ xvi. 22, 23. 330.
„ xx. 8–11. 58.
„ 11. 48.
„ xxiii. 29, 31. 315.
„ xxiv. 10. 16.
„ xxxviii. 26. 312.
Numbers, i. 46. 312.
„ iii. 5–13, 39, 43, 46, 47, 315.
„ x. 11. 161.
„ xiv. 18. 15.
„ xvi. 30. 47.
„ xxii. 22. 282.
„ xxvi. 21. 309.
„ 59. 160, 162, 317.
Leviticus, iv. 11, 12. 311.
„ viii. 3, 4. 310.
„ xxvii. 6. 316.
Deuteronomy, i. 1. 311.
„ 7. 315.
„ ii. 25. 100.
„ iii. 14. 133.
„ v. 1. 311.
„ xi. 24. 315.
„ 25. 100.
„ xxiii. 11–14. 311.
„ xxviii. 26. 68.

INDEX. 383

Texts quoted, &c. continued:—
Deuteronomy, xxxiv. 8. 282.
Joshua, viii. 34, 35. 311.
" xiv. 10. 161.
" xxiv. 33. 161.
Judges, i. 21. 133.
1 Samuel, iii. 7. 66.
2 Samuel, v. 6-9. 133.
" viii. —. 315.
1 Kings, ii. 11. 161.
" ii. vii. ix. xi. 173.
" vi. 1. 150, 161.
2 Kings, iii. 4-27. 327.
1 Chronicles, vii. 22-27. 317.
" xxi. 12-15. 282.
2 Chronicles, ix. 26. 315.
" xii. —. 173, 327.
" xx. —. 327.
Job, ix. 9, 10. 306.
" xxvi. 7. 102.
" 14. 288.
" 16. 80.
" xxxvi. 26-28. 306.
" xxxviii. 1, 2. 301.
Psalm viii. 6. 339.
" xviii. 11, 13. 103.
" xix. —. 16.
" 1. 80.
" 4-6. 26.
" xxiv. 1. 80.
" xxxiii. 6. 359.
" 9. 60, 80, 234.
" xxxiv. 7. 284.
" lxv. 8-11. 306.
" lxviii. 17. 284.
" lxxvii. —. 16.
" lxxxix. 9. 80.
" 37. 278.
" xci. 11. 284.
" xciii. 1. 24.
" cii. —. 339.
" ciii. 20. 250, 284.
" civ. 2. 15, 21.
" 3. 16, 103.
" 4. 282.
" 5. 26.
" 10-23. 278.
" 24. 80, 206.
" 29-30. 274.
" cxix. 89-91. 278.
" cxxxvi. 6. 16.
" cxlviii. 1-6. 278, 359.
" 9, 10. 80.
Proverbs, xxx. 30. 68.
Ecclesiastes, i. 4. 26.
" 5. 26.
" 7. 102.

Texts quoted, &c. continued:—
Isaiah, xl. 6, 8. 274.
" 22. 15.
" xlii. 5. 15, 16.
" xliv. 24. 16.
" xlv. 18. 45, 295.
" xlvii. 5. 16.
" liv. 9, 10. 278.
Jeremiah, iv. 23-26. 45.
" v. 22. 80.
" vii. —. 16.
" xxxi. 22. 47.
" 35, 36. 278.
" xxxiii. 20, 21. 278.
Ezekiel, i. 22. 16.
Daniel, vi. 22. 282.
" x. 13, 20, 21. 282.
" xii. 1. 282.
" 3. 16.
Hosea, ii. —. 16.
Amos, ix. 6. 102.
Nahum, i. 3. 306.
Matthew, iii. 5. 311.
" viii. —. 16.
" xviii. 10. 282.
" xix. 45. 128, 340, 341.
" xxiii. 35. 342.
Mark, ii. 27, 28. 127, 340.
" x. 6-8. 341.
Luke, ii. 34-36. 344.
" 36-38. 343.
" iii. 38. 340, 343.
" vi. 38. 16.
" x. 7. 350.
" xvi. 29, 31. 334.
" xvii. 27. 343.
" xx. 37. 334.
" xxii. 43. 283.
John, v. 47, 48. 334.
" viii. 44. 341.
" ix. 4. 56.
Acts, iv. 24. 340.
" v. 19. 283.
" vii. 2. 344.
" 4. 159, 160.
" 22. 174.
" xii. 23. 282.
" xiii. 20, 21. 161.
" xiv. 15. 340.
" xv. 20. 344.
" xvii. 26. 104, 342, 344, 366.
" 34. 346.
Romans, i. 21, 23. 306.
" v. 12. 86, 342.
" vi. 20. 342.
" 23. 342.
" vii. 2. 341.

Texts quoted, &c. continued:—
Romans, viii. 20, 22. 342.
" xvi. 20. 342.
1 Corinthians, vi. 16. 341.
" xi. 3. 342.
" 7. 340.
" 8, 9. 341.
" xiv. 34. 342.
" xv. 21, 22. 342.
" 22. 128.
" 45, 47. 340.
2 Corinthians, iv. 6. 339.
" xi. 3. 341.
Galatians, iii. 17. 159, 160.
Ephesians, iii. 9. 340.
" v. 30, 31. 341.
Colossians, i. 16. 340.
" 17. 284.
" ii. 5. 17.
" iii. 11. 128.
1 Timothy, ii. 13, 14. 341.
" iv. 3. 343.
" v. 18. 350.
2 Timothy, iii. 16, 17. 348.
Hebrews, i, 1. 348.
" 2. 340, 360.
" 3. 284.
" 10. 339.
" 14. 283.
" ii. 7, 8. 339.
" iii. 4. 340.
" iv. 4. 340.
" vi. 7. 339.
" ix. 27. 342.
" xi. 4. 342.
" 5. 340.
" 7. 343.
" 9. 160.
" xii. 24. 343.
James, iii. 9. 340.
1 Peter, iii. 20. 343.
2 Peter, i. 21. 348.
" ii. 5. 343.
" iii. 5. 339.
" 6. 343.
" 7. 344.
1 John, iii. 8. 341.
" 12. 342.
Jude, 11. 342.
Revelation, ii. 7. 341.

Texts quoted, &c. continued:—
Revelation, iv. 6. 16.
" ii. 11. 340.
" vii. 9. 340.
" viii. 7-12. 282.
" x. 6. 340.
" xii. 9. 341.
" xiv. 7. 340.
" xvi. 1. 282.
2 Esdras, vi. 38-59. 28.

Thames, 190.
Theories, Danger of adopting imperfect, 82.
Tierra del Fuego, 186.
Tinière Delta, 185, 200.
Tohu va bohu, 45.
Trench, Archbishop, 280.
Troyon, M. 181, 185.
Turks, 115.
Tycho, 27.
Tylor, Mr. 251.
Tyndall, Professor, 270.
Types of crania, 111.
" Change in, 115.
Uniformity of nature, 275.
Unity of language, 142.
" mankind, 104.
Universal law, 275.
Van Dieman's Land, 252.
Vitality and aquosity, 266.
Wagner, Professor, 111.
Wallace, Mr. 112, 130, 224.
Warington, Mr. 71.
Waste of space, 298.
Wells, 191.
Wey, 190.
Whewell, Rev. Dr. 22, 27, 40, 206, 374.
Whitney, Professor, 154.
Wilkinson, Sir J. G. 179.
Wisdom of Scripture phraseology, 27, 36.
Wilson, Rev. Dr. 359.
Wordsworth, Bishop, 80, 238, 337.
Worsaae, 196.
Yao, 155.
Young, Professor J. 267.
Yugas, 151.
Zoroaster, 358.

www.ingramcontent.com/pod-product-compliance
Lightning Source LLC
Chambersburg PA
CBHW032014220426
43664CB00006B/239